WHELI

BUSINESS S

AND STATISTI

CHETTY

BUSINESS STATISTICS
AND STATISTICAL METHOD.

WHELDON'S
BUSINESS STATISTICS
AND STATISTICAL METHOD

G. L. THIRKETTLE,
B. Com.(Lond.), F.I.S.
Formerly Senior Lecturer in Statistics
The Polytechnic of North London

EIGHTH EDITION

MACDONALD AND EVANS

MACDONALD AND EVANS LTD.
Estover, Plymouth PL6 7PZ

First published 1936
Second edition 1940
Third edition 1945
Reprinted 1945
Reprinted 1946
Reprinted 1947
Reprinted 1948
Reprinted 1951
Reprinted 1952
Reprinted 1953
Reprinted 1955
Fourth edition 1957
Reprinted 1958
Reprinted 1960
Reprinted 1961
Reprinted 1962
Fifth edition 1962
Reprinted 1963
Reprinted 1964
Reprinted (with amendments) 1965
Reprinted 1966
Reprinted 1967
Sixth edition 1968
Reprinted 1969
Seventh edition 1972
ELBS edition of the seventh edition 1972
ELBS edition reprinted 1974
Eighth edition 1976
ELBS edition of the eighth edition 1976
ELBS edition reprinted 1977
Reprinted 1978

ISBN: 0 7121 2331 8

Printed in Great Britain by Richard Clay (The Chaucer Press), Ltd., Bungay, Suffolk

PREFACE TO EIGHTH EDITION

COMPETENT administration, whether in commerce, industry or government, requires a knowledge of statistical method. This fact has become more and more recognised, as is evidenced by the large number of bodies that now include the subject of Statistics in their examinations.

So that this book should continue to serve students in the future, as it has done so well in the past, new material has been added to keep abreast of the ever-increasing requirements of examining bodies.

A purely theoretical approach has been avoided and emphasis has been placed on applications to business problems.

Appendixes provide tables of logarithms, squares, etc., to save the student the necessity of obtaining separate tables.

I thank all those bodies (a list of which appears at the beginning of *Appendix* I) who have kindly allowed me to reproduce questions from their examinations. Any solutions I have given are solely my responsibility.

Acknowledgments are also due to the Controller of H.M.S.O. for permission to reproduce copyright material from a number of government publications.

G. L. THIRKETTLE

London, June 1976

CONTENTS

CHAPTER 1

INTRODUCTION

The meanings of "statistics." This word is used with two quite distinct meanings. It can refer to facts which can be put into a numerical form, as in the phrase "unemployment statistics." This is the meaning the man in the street gives to the word. It can also refer to statistical methods, which is the subject-matter of this book. Statistical methods are, to use Bowley's description, devices for abbreviating and classifying numerical statements of facts in any department of inquiry and making clear the relations existing between them.

When used with the first meaning, "statistics" is a plural noun, and the statistician usually uses the word data instead. When "statistics" refers to the science it is a singular noun, and this is the meaning usually given to it by the statistician.

The nature of statistical data. Statistics can only deal with numerical data. Often, however, data which are of a qualitative nature can be put into a quantitative form. Health can be measured by the number of days' illness; intelligence, as is well known, by specially designed tests. There is a considerable amount of arbitrariness about this, but it does allow statistical methods to be used, and thus provides additional evidence not otherwise available in given inquiries.

Sometimes the material of an inquiry can be divided up into "cases" possessing a certain attribute and those not possessing it. It is often possible to put qualities in order of rank. Thus, in the case of colours, the lightest could be given rank 1, the next lightest rank 2 and so on. In business, however, most data can be measured or counted directly. Examples would be the number of absentees, sales, quantity produced, wages, etc.

The importance of definition. Before any data are collected it is absolutely necessary to define clearly and unambiguously all the terms used and every piece of information required. Failure to do so will mean that the conclusions drawn from the inquiry will not be valid, and that comparisons over time and between areas will not be true comparisons. The conclusions drawn will be inaccurate because answers to questions will not be answers to the same questions, but only apparently so, the questions having many interpretations unless all terms are clearly defined.

Statistical inquiries. The business man, and indeed anyone who has to administer any organisation, is concerned with inquiries of many kinds. Some of these are capable of being treated statistically, and statistical evidence can be provided in respect of the information wanted.

The steps in a statistical inquiry are as follows:

1. *The problem must be clearly stated.* Suppose the problem concerns wages in a factory. Is it about wages earned or wage rates? Must the statistics concern all employees, or separate grades, both men and women? Should lost time, overtime, piece-work and bonus payments be included or allowed for? Should receipts in kind be included? The purpose of the investigation will provide guidance as to the exact information to be obtained.

2. *Selection of the sample.* If complete coverage of the information available is not made, then the size of the sample and method of sampling will have to be determined. This will depend on the kind of information wanted, the cost and the degree of accuracy required. The best example of a sample inquiry in business is market research.

3. *Drafting the questionnaire.* This is quite a difficult job if the answers obtained are to be of value. Usually a number of questions have to be redrafted to get the exact information wanted. A pilot survey is useful to enable a satisfactory questionnaire to be obtained.

A great deal of information in business, however, is already available in the form of accounting records, costings, administrative information about personnel and so on; questionnaires apart from market research are therefore used only for special inquiries.

4. *Collection of data.* Where not available as administrative records or published, the most satisfactory way is by means of enumerators. Enumerators ask the questions and fill in the questionnaires.

5. *Editing the schedules.* Questionnaires require to be checked, sometimes coded, and calculations made before tabulation can be done.

6. *Organisation of data.* The items require to be counted or the values summed either in total or in various categories before they can be tabulated.

7. *Analysis and interpretation.* Before the information acquired can be used it is analysed and then interpreted. This requires a sound knowledge of statistical methods and also, and this is often lost sight of, a sound knowledge of the subject for which statistical evidence has been obtained.

8. *Presentation.* This may take the form of tables, charts and graphs.

9. *The writing of the report.* This will give the results of the investigation and, where required, will make recommendations. Tables and charts usually play an important part in business reports.

Statistics as a tool of management. In order that the student at the beginning of his studies shall have some idea of the value of statistics to the business man, some problems, which are the concern of management and for which statistical methods are appropriate, are given below.

Stock control. Carrying too large a stock means idle capital and unnecessary costs of storage. Too small a stock means that materials are not available when required, resulting perhaps in lost sales. The "right" amount is a matter, therefore, of considerable importance.

Market research. The business man, in order to be successful, requires to sell what his customers want in the right types and qualities. He wants to know what to sell, in what quantities, when and where.

Quality control. If he produces a mass-produced article, the business man can usually check the quality of his output more efficiently by means of statistical methods.

Sales trends. This is vital information for future planning. Statistical methods are preferable to optimistic guesses, usually called judgment.

The relationship between costs and methods of production. Unless such information is forthcoming, the most efficient method of manufacture will probably not be used.

This matter of statistics and business management is returned to in Chapter 29. Before that, statistical method is dealt with in some detail. The student must master the methods before he can apply them.

Questions

1. "Statistics is the science of counting." Criticise this definition. If you are not satisfied with it, state your own.

2. Why is definition important in any statistical investigation?

3. (i) "The more facts one has, the better the judgment one can make."
   (ii) "The more facts one has, the easier it is to put them together wrong."

Comment on these two views in relation to the principle that statistics are an aid to, not a substitute for, business judgment.

Chartered Institute of Transport.

4. Enumerate the main steps in undertaking a statistical inquiry.

5. In what practical ways can statistics be of service to a business man?

CHAPTER 2

COLLECTION OF DATA

Primary and secondary data. Data may be expressly collected for a specific purpose. Such data are known as primary data. The collection of facts and figures relating to the population in the census provides primary data. The great advantage of such data is that the exact information wanted is obtained. Terms are carefully defined so that, as far as humanly possible, misunderstanding is avoided.

Often, however, data collected for some other purpose, frequently for administrative reasons, may be used. Such data are known as secondary data. Details of imports and exports are compiled by the Statistical Office and Customs and Excise from declarations made by importers and exporters to the local Customs and Excise Officers. This material is obtained for administrative reasons. It is used, however, for compiling quite a large number of statistics relating to overseas trade, including, for example, import and export price indices. The Retail Price Index, on the other hand, uses primary data, the officials of the Department of Employment collecting retail prices expressly for the Retail Price Index each month.

Secondary data must be used with great care. Such data may not give the exact kind of information wanted, and the data may not be in the most suitable form. Great attention must be paid to the precise coverage of all information in the form of secondary data.

Much of the business data used in compiling business statistics is also secondary data, the source often being the accounting, costing, sales and other records. Such data will often require to be adjusted. In finding a sales trend, for example, it may be necessary to adjust the sales figures as given by the accountant for the varying days in the month. Other secondary data used by the business man are published statistics. Before any such material can be used with safety it will be necessary to know the source of the figures, how they were obtained, exact definitions and methods of compilation.

Methods of collecting data. Business data are often collected in the normal course of administration, and not specifically for statistical purposes. However, there is no reason why records should not serve the two purposes, and in such cases care should be taken to ensure that the record is adequate statistically as well as administratively.

The following list covers the most important methods of collecting data:

1. *Postal questionnaire.* This takes the form of a list of questions sent by post. Unless, however, the respondent (the person who is required to answer the questions) has an interest in answering it or is under legal compulsion to do so, the postal questionnaire is generally unsatisfactory, producing few replies, and those of a biased nature.

The postal questionnaire is satisfactory when the law compels the respondent to reply. The Statistics of Trade Act, 1947 makes it compulsory for firms to answer the questionnaire sent out in respect of the census of production.

The postal questionnaire is also satisfactory when sent by trade associations to their members, since the members have an interest in answering it.

Some firms have tried to get answers by offering small gifts. This is not a very good idea, since it will produce biased answers, because the respondent tries to please the donor.

2. *Questionnaires to be filled in by enumerators.* This is the most satisfactory method. The enumerators or field-workers can be briefed so that they understand exactly what the questions mean; they can get the "right" answers; and they fill in the questionnaires more accurately than would the respondents themselves.

In the case of the Census of Population of the U.K., the enumerators do not fill in the schedules. This, however, is quite exceptional.

3. *Telephone.* Asking questions by telephone is not usually a very good method, because people who possess telephones form a biased sample. Telephone interviews are, however, useful for certain kinds of radio research.

4. *Observation.* This method entails sending observers to record what actually happens while it is happening. An example where this method is suitable is in the case of traffic censuses.

Actual measurements or counting also come under the heading of observations. Examples occur in Statistical Quality Control.

5. *Reports.* These may be based on observations or informal conversations. They are usually incomplete and biased, but in certain cases may be useful.

6. *Results of experiments.* This method is of greater interest to the production engineer, the agronomist and other applied scientists than to the business man.

Statistical units. Since the compilation of statistics necessarily entails counting or measurement, it is very important to define the statistical unit which is to be counted or measured.

The definition of the unit is not always as simple as would at first appear. A little reflection on the following words will reveal the need for precise specification in each case: prices may be wholesale, retail, delivered or ex-works; accident may refer to a slight or serious injury, one officially reported or one resulting in a compensation claim; wages may refer to earnings or wage rates.

The requirements for a statistical unit are as follows:

1. *It must be specific and unmistakable.*

2. *It must be homogeneous.* This uniformity is essential; the unit must not imply different characteristics at different times and places. If the selected unit is not applicable to all cases coming under review, it is often possible to overcome the difficulty (*a*) by sub-dividing the data into groups or classes until sufficient uniformity has been secured, (*b*) by expressing dissimilar units in terms of equivalents of the selected unit. *E.g.*, if the output of bakeries for a period were being compared, some producing 1-, 2- and 4-lb loaves, all sizes could be reduced to the equivalent of, say, 2-lb loaves.

3. *It should be stable.* If it is desired to use a fluctuating unit, such as a calendar month, then adjustments will require to be made before comparisons will be valid.

4. *It must be appropriate to the inquiry and capable of correct ascertainment.* When compiling labour statistics it is necessary to select the unit appropriate to the information required—*e.g.*, workers engaged directly on production, those employed in indirect factory services, those in administrative offices or those in the sales department.

Types of statistical unit. A simple statistical unit—*e.g.* metres, litres, £ sterling—are not difficult to define, but care must be used in some cases. For example, a ream may be 480 sheets, but often consists of 500 or 516 sheets. A ton may be 2,240 lb or it may be a short ton of 2,000 lb, or again it may be a metric ton.

A composite unit may have to be used in some cases. Thus, electric power is measured in units of kilowatt-hours. Railway transport is measured in ton-miles (*i.e.* number of tons × number of miles carried) or in passenger-miles. Education authorities measure the amount of instruction given in student-hours.

The form of questionnaire. The questionnaire is often in two parts. The first part is a classification section. This requires such details of the respondent as sex, age, marital status, occupation. The second part has the questions relating to the subject-matter of the inquiry. The answers given in the second part can be analysed according to the information in the first part. There will also be questions of a purely administrative nature, such as the date of the interview, the name of the interviewer, etc. Market research and social survey questionnaires are usually of this form.

Concerning the layout of questionnaires, there is a tendency, where it is possible, to provide all the answers that could be given to a question on the form, and the appropriate answer is circled or ticked. This treatment can only, of course, apply to certain questions.

The characteristics of a good questionnaire. The list of desirable qualities that a questionnaire should possess given below would seem to be a matter of common sense. Nevertheless, the drafting of questionnaires is one of the most difficult tasks of an inquiry. A pilot survey—that is, a trial survey, carried out prior to the actual survey—invariably leads to alterations and improvements in the questionnaires.

1. *Questions should not be ambiguous.* This means that the question must be capable of only one interpretation.
2. *Questions must be easily understood.* Technical terms should be avoided, except where the questionnaire is addressed to specialists.
3. *Questions should be capable of having a precise answer.* The answer should take the form of "yes" or "no," a number, a measurement, a quantity, a date, a place; facts are required, not opinions (except where opinions are wanted, as in opinion polls).
4. *Questions must not contain words of vague meaning.* To ask if something is large or if a man is unskilled are examples of such questions. When does something become large? What jobs are unskilled?
5. *Questions should not require calculations to be made.* Such questions give rise to unnecessary sources of error. If for the purpose of the inquiry it is necessary to know annual earnings, but the respondent is paid weekly, the weekly earnings are asked for. The calculations necessary are done by the statistician's staff.
6. *Questions should not require the respondent to decide upon classification.*
7. *Questions must not be in such a form that the answers will be biased.* The questions will not therefore contain emotionally coloured words, they will not be leading questions—that is, they will not put answers into the respondents' mouths—and, of course, they must not give offence.
8. *The questionnaire should not be too long.* If a questionnaire is too long, the respondent will not be co-operative, and this may mean inaccurate answers; often a way out is for some of the respondents to answer part of the questions and other respondents to answer other questions.
9. *The questionnaire should cover the exact object of the inquiry.* However, provided the questionnaire is not made too long, advantage can be taken of the arrangements made to obtain information

Board of Trade

CENSUS OF DISTRIBUTION AND OTHER SERVICES
1950

Confidential Return

20
Service Trades

	Machine Code
Your return should cover the whole of your business carried on at this address, including the sale of goods and any catering which you undertake. Further notes for guidance are given on a separate sheet.	
1. If you cannot give figures for a full year because you began business during 1950, you should give figures for that part of 1950 in which business was carried on and state here the period: From........................ To........................	A.20.1
2. Name and Address Trading name .... Address of shop, studio, order office, etc. .... Owner's name	
Does the owner carry on a similar business at any other address? ....(yes or no)	A.19.2
If "yes," give the address of the owner (where it differs from the above) and send the form immediately to that address for completion. Owner: See Note 2 on attached sheet. (Address).. .... Is any business conducted on your premises by other traders, for example, a "shop within a shop"? ....(yes or no)	
3. Description of Business *a.* Give the usual description of your business, for example, hairdresser, portrait photographer, street photographer, undertaker	A.23.2
b. If your business includes that of a hairdresser, put a cross in the square against the kind of service normally provided: ☐ Men's hairdressing ☐ Ladies' hairdressing ☐ Beauty treatment	
6. Purchases State the net cost to you, including purchase tax, of goods and materials purchased during the year for business purposes: £.................. (Include payments to other firms for work given out and fuel and electricity, but exclude capital items such as new vehicles, machinery, fittings and equipment)	A.41.10
7. Stocks What was the total of your stocks of goods and materials (including fuel) on hand, for sale, or for use in your business, at cost or market value whichever is the lower	
a. at the beginning of the year? £..................	S.31.7
b. at the end of the year? £..................	S.41.7

relating to a different inquiry. In the last population census of the U.K., questions were asked about sanitary arrangements.

When drafting questionnaires, it is a good plan to check the questions against the above list.

Interviewing. The enumerator does not usually deliver the questionnaires, but, instead, asks the respondent the questions listed. He, or she (and the majority of enumerators are women), also fills in the answers. Upon the enumerator, then, falls the duty of obtaining accurate information. The interviewer must not express his own opinions, nor must he allow them to influence the answers. Complete objectivity must be observed. The interviewer must merely record facts accurately—that is, the answers as given.

The enumerator should first make sure he is interviewing the right person. He should then explain the purpose of the interview as clearly and as briefly as possible, soliciting the help of the respondent. If help is refused, further details about the inquiry may succeed in obtaining the required co-operation.

If the respondent does not understand the question, or has not given a clear answer, resort is made to "probing." This consists of re-wording or explaining the question, or sometimes asking other questions to help a person to remember. It also consists of asking a respondent to explain his answers more fully. Great care must be taken not to put answers into the respondent's mouth. Where the question asks for an opinion, all the probing allowed to the enumerator is to ask the respondent to explain an unclear answer more fully. He must not explain or re-word questions.

Questions

1. Give some account of the various methods which can be used in the collection of statistical data. *Institute of Company Accountants.*

2. "The collection of accurate statistical data upon a large scale is only possible with exact definition of terms, *e.g.* units, classes, etc." Comment with illustrations from the transport industry.
Chartered Institute of Transport.

3. Design a questionnaire to investigate the amount spent on clothing and footwear throughout the country. State the precautions you would take to avoid any bias in your results.
London Chamber of Commerce and Industry.

4. By drawing on your own or related experience in the transport industry, illustrate the need for, and importance of, exact definition for statistical quantities. *Chartered Institute of Transport.*

5. List the main points to be observed when drafting a questionnaire.
Institute of Company Accountants.

6. Name four ways in which statistics can be of service to a manufacturer.

7. Distinguish between primary and secondary data, giving examples of each.

8. What care must be taken when using secondary data?

9. What is meant by probing? When is it permissible?

CHAPTER 3

SAMPLING

The sampling process. Instead of obtaining data from the whole of the material being investigated, sampling methods are often used in which only a sample selected from the whole is dealt with, and from this sample conclusions are drawn relating to the whole. If the conclusions are to be valid, the sample must be representative of the whole. The selection of this sample must therefore be made with great care.

The advantages of using sampling methods. There are a number of advantages to be gained by using sampling methods. Among these may be listed the following:

1. *Quick results.* The results can be obtained more quickly, for two reasons: the data are obtained more rapidly and can be analysed more quickly. Speed in analysing the data is also possible by analysing a sample of the data before analysing the whole. The analysis of a 1 per cent sample of the 1951 population census was such an example.

2. *Higher quality interviewers.* When the respondents constitute only a small proportion of the population they are more willing to give detailed information, and because fewer interviewers are required, a much higher quality of interviewer can be employed.

3. *More skilled analysis.* Higher-grade labour can be employed on the computing, tabulating and analysis of the data.

4. *Lower costs.*

5. *Following up non-response is much easier.*

6. *The error can be assessed.*

7. *Guarding against incomplete and inaccurate returns is easier.* A sample is often used as a check on the accuracy of complete censuses.

8. *Practical method.* When the investigation entails the destruction of the material—*e.g.* to discover the life of a lamp—sampling methods are the only practical methods to use.

The "population" or "universe." These terms refer to the whole of the material from which the sample is taken. The frame will consist of all the items in the population, or some means of identifying any particular item in the population. This frame is necessary so that any

A sampling frame is a list of all the members of the population from which the sample is selected item in the population can be part of the sample. The frame must be complete—that is, no item of the population should be left out—and it should not be defective because of being out of date, or contain inaccurate, or duplicate items, or inadequate because it does not cover all the categories required to be included in the investigation.

Examples of frames suitable for certain inquiries are the electoral lists and the Jurors Index. For the sampling of dwellings in built-up areas, town maps provide a useful frame.

Sometimes, no suitable frame is available, the population required may be part of a larger population, and cannot be identified within it. One method that can then be used is to make the first question determine if the respondent is within the special population required. This method entails a good deal of work to get the necessary sample.

The basic statistical laws. The possibility of reaching valid conclusions concerning a population from a sample is based on two general laws, namely, the law of statistical regularity and the law of the inertia of large numbers.

1. *The law of statistical regularity*. The law of statistical regularity states that a reasonably large number of items selected at random from a large group of items will, on the average, be representative of the characteristics of the large group (or "population").

The important features are:

(*a*) The selection of samples must be made at random, as for instance by some lottery method, so that every item in the population has a chance of being in the sample.

(*b*) The number of items in the sample must be large enough to avoid the undue influence on the average of abnormal items. The larger the number of items selected, the more reliable is the information afforded. The sampling error is inversely proportional to the square root of the number of items in the sample.

2. *The law of inertia of large numbers.* The law of the inertia of large numbers is a corollary of the law just described. It states that large groups or aggregates of data show a higher degree of stability than small ones.

The movements of all the separate components of the aggregate reveal a tendency to compensate one another, some probably moving higher, others lower; but, taking a large number of data, it is unlikely they will all move in the same direction. The greater the number composing the aggregate, the greater will be the compensation, or tendency of movements to neutralise one another, and consequently the more stable will be the aggregate.

Sampling errors. Even when a sample is chosen in a correct manner, it cannot be exactly representative of the population from which it is chosen, for not all samples will be alike. This gives rise to sampling errors.

The amount of the random sampling error will depend on the size of the sample—the larger the sample, the smaller the error—the sampling procedure involved, and the extent to which the material varies.

In the case of sampling methods which are based on random sampling, it is possible to measure the probability of errors of any given size. By error is meant, of course, the difference between the estimate of a value as obtained from the sample and the actual value.

Errors due to bias. In addition to random sampling errors, errors due to bias may arise. Unlike sampling errors, errors due to bias increase with the size of the sample. Furthermore, it is not possible to measure the amount. Hence, every care must be taken to avoid these kinds of errors. They may arise in a number of ways:

1. *Deliberate selection.* Whenever the personal element enters into the selection, bias is inevitably introduced.
2. *Substitution.* Substituting another item for one already chosen in the sample will also cause bias. This may happen if, for example, a person chosen for interviewing is out, and someone else is then interviewed. People who are at home all day have not the same characteristics as those who spend most of their time away from home.
3. *Failure to cover the whole of the sample.* Even if no attempt is made to substitute and the whole sample is not covered, there will be bias. This is a frequent fault with postal questionnaires, when not all the respondents will reply.
4. *Haphazard selection.*

Random sampling. The word random does not mean haphazard. It refers to a definite method of selection. In a *simple random sample* each unit of the population has exactly the same chance as any other unit of being included in the sample. One method of obtaining a random sample, not a very practical one, would be to number or name every unit of the population, and to place slips, each bearing a number or a name, one slip for each unit of the whole of the population, in a drum, as in a lottery, and, after thoroughly mixing them, draw from the drum the number of slips required to obtain the required size of sample. In practice, a table of random numbers is used. A small portion of such a table is shown in Fig. 1. These numbers can be read in any methodical way: vertically, horizontally, diagonally, forwards or backwards, etc. Any number chosen must be capable of being assigned unambiguously to a particular unit of the population.

Suppose the population to consist of 999,999 items, the sample required 1,000, then the first four items to be chosen in the sample might be, using the table shown, numbered 162,277; 844,217; 630,163; 332,112. The figures have been read across for each number, but downwards to get the next number, starting with the second group of numbers, downwards, on the left. Countless alternatives could have been chosen.

03	47	43	73	86	36	96	47	36	61	46	98	63	71	62
97	74	24	67	62	42	81	14	57	20	42	53	32	37	32
16	76	62	27	66	56	50	26	71	07	32	90	79	78	53
12	56	85	99	26	96	96	68	27	31	05	03	72	93	15
55	59	56	35	64	38	54	82	46	22	31	62	43	09	90
16	22	77	94	39	49	54	43	54	82	17	37	93	23	78
84	42	17	53	31	57	24	55	06	88	77	04	74	47	67
63	01	63	78	59	16	95	55	67	19	98	10	50	71	75
33	21	12	34	29	78	64	56	07	82	52	42	07	44	38
57	60	86	32	44	09	47	27	96	54	49	17	46	09	62
18	18	07	92	46	44	17	16	58	09	79	83	86	19	62
26	62	38	97	75	84	16	07	44	99	83	11	46	32	24
23	42	40	64	74	82	97	77	77	81	07	45	32	14	08
52	36	28	19	95	50	92	26	11	97	00	56	76	31	38
37	85	94	35	12	83	39	50	08	30	42	34	07	96	88
70	29	17	12	13	40	33	20	38	26	13	89	51	03	74
56	62	18	37	35	96	83	50	87	75	97	12	25	93	47
99	49	57	22	77	88	42	95	45	72	16	64	36	16	00
16	08	15	04	72	33	27	14	34	09	45	59	34	68	49
31	16	93	32	43	50	27	89	87	19	20	15	37	00	49

Fig. 1.—Small portion of a table of random numbers.

Stratified sampling. Where the population is heterogeneous, a stratified sample is required. This will increase the accuracy of the results, provided the strata relevant to the investigation are chosen. The population is divided into strata, blocks of units, in such a way that each block is as homogeneous as possible. Each block or stratum is then sampled at random. If the same proportion of each stratum is taken, each stratum will be represented in the correct proportion in the sample. This eliminates differences between strata from the sampling error.

Sometimes instead of a *uniform sampling fraction*, *variable sampling fractions* are employed. The more variable strata in the population is more intensively sampled. This will increase the precision.

Systematic sampling. When a list of all the units of a population is available a sample of every *n*th item would be known as a systematic sample. The first entry would be obtained by random selection. However, unless the list is arranged at "random" (in the technical sense), such systematic sampling is not random sampling.

Provided there are no characteristics of the items in the list which occur periodically with the same interval as the sampling interval, the sample will not be biased. It is also possible to obtain an estimate of the sampling error good enough for most practical purposes.

Multi-stage sampling. This is best explained by an example. Instead of obtaining a sample from, say, all the households in the country, a random sample is obtained of all the rating authorities. A random sample would then be chosen of the households in each of the areas of the rating authorities chosen at the first stage of sampling.

Quota sampling. The sample is divided up into quotas, the quotas indicating the number of people to be interviewed, but leaving the choice of the actual respondents to the interviewers. This, of course, introduces bias. The quotas are chosen so that the sample is representative of the population in a number of respects, according to the controls chosen. It may be completely unrepresentative in other respects. Suppose a sample of 400 is required, that the controls chosen are sex, and whether householder or not. Let the proportion of men to women in the population from which the sample is chosen be 9 to 11, and the proportion of householders to non-householders be 3 to 2. The quotas would be as follows: 132 householders who are women, 88 women non-householders, 108 householders who are men and 72 men non-householders. The sample of people interviewed might be completely unrepresentative as regards age. But this might not be relevant for the inquiry. It is not possible to measure the sampling error.

Questions

1. What is a statistical sample? When and why are samples used? Discuss the essentials of the various methods of sampling.
Institute of Company Accountants.

2. Describe a method of drawing a random sample for a postal survey of annual purchases of toothpaste by households in the Southern Counties. Design a short questionnaire, including any necessary instructions for completion. *Institute of Statisticians.*

3. Explain the following:
Law of statistical regularity.
Law of inertia of large numbers.
What section of statistical method depends on these two laws?
Institute of Company Accountants.

4. Explain the use of sampling in statistical investigations. What considerations should principally be borne in mind when making a social inquiry by sample?
Local Government Examinations Board.

5. What is the object of sampling? Illustrate it by reference to actual cases of which you are aware, referring to some different kinds of sampling and the dangers to be avoided. *Institute of Cost and Management Accountants.*

6. Describe the sample method of investigation. Illustrate your answer by reference to any official or non-official statistical investigation with which you are familiar.

7. Describe three of the more important methods of sampling used in conjunction with statistical surveys.

Suggest methods which may be employed in statistical surveys to validate the sample information obtained. *I.C.S.A. Part 1, June 1975.*

8. What is a statistical sample? Discuss the essentials of the various methods of sampling and state and explain the two laws upon which the sampling technique depends. *Institute of Company Accountants.*

9. Explain carefully the meaning of the following terms:

- **(*a*) population;**
- **(*b*) frame;**
- **(*c*) random sample;**
- **(*d*) quota sampling;**
- **(*e*) sampling error.**

10. In what ways can a faulty selection of the sample give rise to bias?

CHAPTER 4

SURVEYS

A survey consists in finding facts in particular fields of inquiry. In this chapter, a short description is given of three important types of surveys in which the data collected are of a statistical nature.

Social surveys. Although many bodies conduct social surveys—as, for example, the universities—the most important are the social surveys of the Government.

The social survey came into being to provide information for other Government departments so that they could carry out their duties more efficiently. Government departments often have to deal with special social problems. The surveys help the Government committees to arrive at a policy for solving the particular problems. A large part of their work consists in providing continuous statistics, such as the Consumer Expenditure inquiries which are published in the *National Income and Expenditure Blue Books*.

The survey is a sample survey, the size of the sample depending on the permissible sampling error and the amount of analysis required (it does *not* depend on the size of the population). The sample may therefore vary from 400 to 4,000, or even more. The sampling is always some form of "random" sampling, the word "random" being used in its technical sense (see Chapter 3).

The actual method used is usually some form of multi-stage sampling —*e.g.* a sample of the 1,470 administrative districts of England and Wales and the fifty-seven administrative districts of Scotland is first taken, and from each of the districts chosen a sample of people will be taken from the card-indexes kept in each of the administrative districts. In the case where a sample of householders is required, the rating records kept by local authorities are used. Quota sampling is used only on some pilot surveys when a fully representative sample is not so important, the aim of the pilot survey being to establish the most satisfactory form of the questions to be asked; random sampling is used for the actual survey.

Market research. Since the aim of the producer is to sell his goods, it is necessary for him to know the "right" kind of commodity to produce, and, in order to prevent losses, to know in what quantities and at what price. This entails the producer knowing the consumers' needs and desires, and if he attempts to persuade them of the superiority of his particular products, knowing the results of his publicity and his competitors' successes. This part of market research lends itself admirably to statistical methods.

As an example of this kind of research, a brief description is given of how the size of audiences, the potential customers of the advertisers, are measured in the case of commercial television.

The following methods are possible:

1. *Telephoning a sample of viewers*. They are asked what they are viewing at the time they are called. This method has two main disadvantages: the information obtained is limited to what is being seen at the particular moment of the call, and the sample is not representative, since not every family has a telephone.

2. *Personal interview*. The respondent enumerates the programmes he has seen in a given period and his reaction to them. He is also asked what other members of the family saw the programmes and their reactions. (The B.B.C. use this method.)

3. *The "viewing" diary*. Selected families keep a diary of the programmes they view for, say, a week and answer a number of questions on them. This method (one used in the U.S.A.) will give biased results, as only the more educated are likely to keep the diary in a reliable manner.

4. *Invited audiences*. Representative audiences are invited to view several programmes, after which they are questioned on their reactions to each programme. This method is also used in the U.S.A.

5. *Automatic recorders*. These devices are attached to the television sets of a very carefully selected sample of viewers. They record when the set is on and to what station it is tuned. Market research organisations who use such devices are the Nielsen Company who use the "recordimeter" and the "audimeter" for recording and measuring the amount of viewing, and Television Audience Measurement Ltd., who use the "tammeter."

Two great difficulties face market-research workers. First, no lists are usually available for the relevant populations. There is no list available, for example, of gardeners, which might be useful in the case of market research on gardening implements. Next, random sampling would entail too high a cost. Hence quota sampling is resorted to, with its attendant bias and a sampling error which is not measurable. However, with properly trained interviewers and care it is possible to get useful results.

Public opinion polls. The most widely known organisation in this country which carries out polls on public opinion is the British

Institute of Public Opinion (B.I.P.O.). It is commonly known as the Gallup Poll, and is one of a number of similar independent institutes throughout the world, all affiliated to the Association of Public Opinion Institutes.

Both random and quota sampling are used. Random samples are used in individual constituencies in the case of election polls and are based on the electoral register. Quota sampling is used nationally. The controls for quota sampling are: (*a*) regional, (*b*) rural and urban, in the proportion of 1 to 4, (*c*) size of town, (*d*) sex of respondent, (*e*) age group, (*f*) socio-economic group, four divisions, (*g*) political party of sitting member in the constituency. Some 2,000 people are interviewed.

The B.I.P.O. does not only measure political opinions. Among the data accumulated by the Institute are such facts as: the average English woman is 5 feet 4 inches tall, weighs 9 stone 4½ lb., has brown hair and blue eyes; one woman in three would prefer to be a man; and that during the course of a year nearly one in two spend something on the pools.

Questions

1. What are the characteristics of a good questionnaire to be used by interviewers in an opinion survey? *Institute of Statisticians.*

2. Draft a short questionnaire for a pilot inquiry into householders' preferences in floor polishes.

Suggest the types of alterations which you may have to make to the form for the full-scale inquiry when the results of this pilot survey have been scrutinised. *Institute of Statisticians.*

3. A large transport organisation has asked you to undertake a survey of the travel habits of the population in its area with a view to the better administration of the transport services.

Draft a plan of action to carry out this survey stating what information will be essential; give also some suggestions for further information which you consider might be helpful. Which method or methods would you use to obtain your data? *Institute of Company Accountants.*

CHAPTER 5

PREPARATION FOR TABULATION

Editing the data. Editing the recorded data will often be necessary before proceeding to tabulation. This will, in fact, always be the case when dealing with questionnaires.

The process of editing will consist of:

1. *Checking that the schedules are apparently correctly completed.*
2. *Ensuring that there are no inconsistent entries.*
3. *Entering any figures that have to be calculated.* (Respondents should never be required to make calculations.)
4. *Coding the answers, when necessary.*

Analysis of data. The data can be analysed directly from the forms or record cards. In this case tally-sheets are often useful. Squared paper is an aid to neat and hence accurate work. When counting numerous sorted data, it is usual to mark strokes on the working paper in fives arranged as in Fig. 2.

Groups of five are very easy to count.

Often the data are transferred to cards so that the sorting can be done more easily. In this case, coding will be necessary. The cards may be ordinary cards, in which case sorting them into categories will be done manually; with Cope-Chat cards the sorting is done by means of a needle, and in the case of punched cards mechanically.

In some cases, the original record card can also be used as the sorting card. Such cards are known as Dual-purpose cards. An example is shown in Fig. 3.

Coding. Before the information on the schedules or questionnaires is transferred on to cards for sorting, the information must be coded. Coding consists of giving a code number to each classification. Items can then be sorted under code numbers. Examples of coding are as follows:

(*a*) To designate sex, two code numbers would be required, 1 and 2.

(*b*) If it were required to classify incomes in four groups, (1) under £800 p.a., (2) £800 but under £1,400, (3) £1,400 but under £2,000 p.a., and (4) £2,000 and over, four code numbers would be required —1, 2, 3 and 4.

Numbers, except when they are to be grouped, do not require to be coded.

Cope-Chat cards. Specimen Paramount cards, manufactured by Copeland-Chatterton Co., Ltd., are shown in Fig. 3. These cards are made in a number of suitable sizes. They have holes along each edge.

NUMBER OF WAGE-EARNERS AT VARIOUS HOURLY RATES

							TOTAL
60p-79p	𝍸	𝍸	𝍸	|			16
80p-99p	𝍸	𝍸	𝍸	𝍸	|||		23
£1·00-£1·19	𝍸	𝍸	𝍸	𝍸	𝍸	||||	29
£1·20-£1·39	𝍸	𝍸	𝍸	||			17

Fig. 2.—Counting sorted data.

On the body of the card can be recorded information if they are to be used as a dual-purpose card. A punch is used to make the round holes into V-shaped notches. They can then be sorted by means of a needle, as shown in Fig. 3. The cards are put in a pack (and to ensure they are the correct way round one corner is cut off), a needle is inserted in the

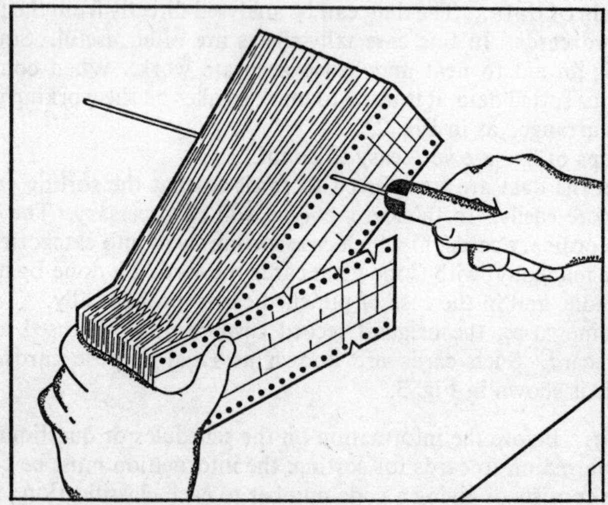

Fig. 3.—Sorting marginally punched cards.

requisite hole and the pack is then lifted and very thoroughly shaken. The cards having the notch fall out. These are the cards recording the information punched.

Since the cards have only a limited number of hole positions, it is necessary to make the utmost use of them. Where it is necessary to make provision for numbers from 1 to 9 to be recorded, four holes can be used, as follows. Let the holes denote values 1, 2, 4 and 7. The other numbers can be coded by using two holes—*e.g.* to denote 6, holes 2 and 4 would be punched.

These cards have one advantage over other cards, even over those used in the more elaborate systems. It is possible to obtain any particular card more quickly than by any other method, since no elaborate sorter is required.

Questions

1. Describe briefly the design and uses of Cope-Chat or any other similar system of small hand-punched cards.

Discuss the relative advantages of different methods of coding numerical information on such cards. *Institute of Statisticians.*

2. Give a short account of the way in which questionnaires are edited.

3. The following is an extract from one of a large number of reports on road accidents collected in 1975. You have been asked to prepare a statistical report on road accidents, and intend to make the analysis with the help of edge-punched cards, on which there are only 40 holes. Design a master card for coding as much relevant information as possible.

Time of day	10.15 a.m.
Vehicles involved	Bus and bicycle
Number of persons injured	1 dead, 2 seriously injured, 1 slightly injured
Weather	Rain
Distance from railway station	2 miles
In speed limit area	No
Class of road	A
State of road surface	Bad
Total number of persons involved	43
Number of witnesses	5
Standard of street lighting	Good

Institute of Statisticians.

CHAPTER 6
TABULATION

Data, however obtained, must be tabulated or put into tables before being used for analysis. Professor Bowley refers to tabulation as "the intermediate process between the accumulation of data, in whatever form they are obtained, and the final reasoned account of the results shown by the statistics."

It is quite obvious that the layout of tables is dependent on the information to be presented and the purpose for which they are prepared. Nevertheless, there are a number of principles which underlie the preparation of all tables. These will be considered in this chapter.

Informative or classifying tables. These are original tables which contain systematically arranged data compiled for record and further use, without any intention of presenting comparisons, relationships or the significance of the figures. In other words, they merely provide a convenient means of compiling data in a form for easy reference, very frequently in chronological order. This type of table is frequently referred to as a schedule.

General or reference tables. These tables contain a great deal of summarised information. They are not used for analytical purposes and usually do not give averages, ratios or other computed measures. If they are embodied in a report, they are usually relegated to an appendix. They are often the source from which summary tables are compiled.

Text or summary tables. These tables are used to analyse or to assist in the analysis of classified data. They show only the relevant data of the question being discussed. Ratios, percentages, averages and other computed measures are often added. If included in a report, they are found in the body of the text. These tables are interpretative or derivative tables, in that they are analytical and are prepared to present significant aspects of the data.

Simple and complex tables. A simple table presents the number or measurement of a single set of items having the characteristics stated at the head of a column or row which forms the basis of the table. Using the terminology of statistics we may tabulate a dependent variable (number of families) against the independent variable (amount per head for food), as in Fig. 4.

A complex table presents the number or measurements of more than one group of items set out in additional columns or rows, and the table is often divided into sections. Such tables generally show the

Food Expenditure of Families in Receipt of Wages

Amount per head for food	Number of families
£1·00 to £1·99	58
£2·00 to £2·99	96
£3·00 to £3·99	82
£4·00 to £5·00	28
	264

Fig. 4.—A simple table.

relationship of one set of data to another or others, and are often arranged so that comparisons may be made between related facts. An example of a complex table is given in Fig. 7. It is rather a full table, and the information might be made more effective by dividing it into two or even more separate tables.

In the case of business statistics it is more usual to find complex tables, because to facilitate a proper consideration of all related facts full information has to be included. Comparative figures, whether absolute or percentages or averages of various kinds, are frequently required, and are therefore incorporated in many forms of tabulated statistics. In other fields of inquiry, more simple tables are usual.

Frequency distribution tables. These give the number of items of different sizes. The example given of a simple table in Fig. 4 was a frequency distribution table.

Time-series tables. These show the values of a variable over a period of time (see Fig. 5).

Construction of tables. The first step is to prepare a rough draft. In making a suitable layout, it is usually found necessary to alter the original design. The alteration often consists in changing the rows to columns or the other way round. The cardinal principles involved are clarity and ease of reference. A table should fill the space allotted to it. Equal margins should be left on both sides and more space allowed at the bottom than at the top. The headings of the columns and rows should be clear, but concise. A title must be given to the table, which must be clear and bold. Although conciseness is necessary, the title

XYZ, Ltd. : Sales

	£
1967	16,237
1968	17,362
1969	23,117
1970	33,206
1971	35,207

Fig. 5.—A time-series table.

must nevertheless give adequate information. If more than one line is required, the pyramid formation is usual. Sometimes the "pyramid" is inverted thus:

EXPORT OF MACHINERY
UNITED KINGDOM
1960–1970

Columnar layout. Fig. 7 shows a completed table illustrating the principles and practice involved.

Arrangement of data. Data are arranged according to the purpose of the table. This may be according to size or importance, chronologically, geographically, or alphabetically. It may be according to

Employees in XYZ, Ltd.

Age	Operatives			Administration			Total		
	Male	Female	Total	Male	Female	Total	Male	Female	Total
Under 18									
18 and over									
Total									

Fig. 6.—A table in blank for presenting the distribution of employees of a firm according to (a) Age, (b) Sex and (c) Employment.

some basis of classification, or merely some usual order. Data to be compared are usually placed in parallel columns, probably adjacent.

Treatment of units. These must be given fully and accurately—*e.g.* price per metric ton. When a number of columns have the same unit, this is placed over the several columns and not repeated over each (see Fig. 7).

Spacing and Ruling. Clarity, and emphasis, where necessary, can be helped by the judicious use of variations in type, spacing and ruling. Where there are many rows of figures, a break should be made, say,

Total Deliveries of Electronic Computers

£000, current prices

		Digital computing systems		Analogue and hybrid computers		Data transmission equipment		Ind. electronic control equipment		Peripheral equipment		Total deliveries		Repair and maintenance		R&D on customer account		Total output		Goods sold to other UK mfrs.
		Total	Export	Total	Export	Total	Export	Total	Export	Total	Export	Total	Export	Total	Export	Hard-ware	Soft-ware	Total	Export	
1966		78,684	16,572	1,393	307	1,973	572	4,667	377	28,221	15,752	114,938	33,580	5,924	313	326	54	121,242	33,893	4,264
1967		88,771	22,827	1,547	293	3,533	838	2,551	333	30,831	13,011	127,233	37,302	8,574	1,070	466	76	136,349	38,394	8,483
1968		113,266	26,579	1,960	233	5,232	1,144	4,902	1,173	39,987	11,711	165,347	40,840	11,868	707	56	242	177,513	41,547	9,847
1969		119,732	36,493	3,897	518	10,182	2,997	9,404	2,803	64,038	19,507	207,253	62,318	15,767	1,010	49	89	223,158	63,328	11,607
1970		171,319	55,698	2,364	126	20,035	9,036	14,674	5,929	87,780	26,142	296,172	96,931	20,840	61	200	384	317,596	97,113	13,779
1969	Q1	25,409	7,181	604	102	2,050	200	2,386	103	13,719	3,467	44,168	11,053	3,399	229	7	30	47,604	11,282	3,522
	Q2	26,032	7,877	1,684	114	2,370	551	1,829	742	17,151	5,023	49,066	14,307	3,896	385	3	32	52,997	14,692	2,922
	Q3	32,038	11,271	814	198	2,148	663	1,148	400	15,214	5,021	51,362	17,553	4,021	266	19	3	55,405	17,819	2,834
	Q4	36,253	10,164	795	104	3,614	1,583	4,041	1,558	17,954	5,996	62,657	19,405	4,451	130	20	24	67,152	19,535	2,329
1970	Q1	35,808	11,700	179	79	2,995	1,307	2,770	948	21,338	5,844	63,090	19,878	4,981	10	24	100	68,195	19,955	3,547
	Q2	41,078	11,685	291	34	5,028	2,126	3,130	1,178	20,246	6,064	69,773	21,087	4,858	17	22	63	74,716	21,104	2,589
	Q3	49,559	16,479	1,641	6	6,237	2,520	3,993	1,829	23,123	6,121	84,553	26,955	5,130	15	53	84	89,820	26,978	3,339
	Q4	44,874	15,834	253	7	5,775	3,083	4,781	1,974	23,073	8,113	78,756	29,011	5,871	19	101	137	84,865	29,076	4,304
1971	Q1p	32,018	8,593	162	10	3,710	1,363	2,967	377	25,398	9,025	64,255	19,368	6,157	17	135	55	70,602	19,385	3,331
	Q2p	33,765	11,085	6	—	4,403	1,511	3,346	841	20,993	7,949	62,513	21,386	6,504	14	70	135	69,222	21,438	2,518
	Q3p	42,989	14,751	1	—	5,304	1,298	4,143	1,123	25,467	10,458	77,904	27,630	7,918	17	36	157	86,015	27,680	3,256

Source: *Department of Trade and Industry.*

Fig. 7.—An example of complex tabulation.

after every fifth row. In the case of, say, monthly figures, a break after every three months.

Printed and typewritten tables can often be set up without the use of ruling, by proper spacing. For instance, major vertical divisions can be separated by a wide space, and related columns can be set closer together.

Footnotes to tables. These are used for four main purposes:

1. *To point out any change in the basis of arriving at the data.* For example, sales may be recorded as "ex-factory" for some of the entries and at delivered prices for others. Any heterogeneity in data recorded must be disclosed to avoid wrong conclusions being drawn.

2. *Any special circumstances affecting the data.* For example, a fire in the works must be noted.

3. *To clarify anything in the table.*

4. *To give the source.* This must always be done in the case of secondary data.

Questions

1. Construct a blank table in which could be shown the number of fires, both serious and slight, in an industrial town. The table should include:

(*a*) the place of the fire, classified into the groups—factories, warehouses, private dwellings, offices, truck and motor lorries, etc.;

(*b*) the cause of the fire, classified into carelessness, electrical defects and unknown;

(*c*) by whom the fire was discovered and extinguished.

2. Lamps are rejected at several manufacturing stages for different faults. 12,000 glass bulbs are supplied to make 40-watt, 60-watt and 100-watt lamps in the ratio of 1 : 2 : 3. At stage 1, 10% of the 40-watt, 4% of the 60-watt, and 5% of the 100-watt bulbs are broken. At stage 2 about 1% of the remainder of the lamps have broken filaments. At stage 3, 100 100-watt lamps have badly soldered caps and half as many have crooked caps; twice as many 40-watt and 60-watt lamps have these faults. At stage 4 about 3% are rejected for bad type-marking, and 1 in every 100 are broken in the packing which follows.

Arrange this information in concise tabular form. Which type of lamp shows the greatest wastage during manufacture? *Institute of Statisticians.*

3. Design a blank (or skeleton) table to bring out the differences in direct and indirect taxation in the U.K., U.S.A., and France for the two years 1950 and 1970.

Take income tax as direct and other taxes as indirect taxation; subdivide the latter if you think it desirable. Both actual and "per capita" figures should be allowed for. Suggest a suitable title.

Institute of Company Accountants.

4. Construct a blank table in which could be shown, at two different dates and in five industries; the average wages of the four groups, males and females, eighteen years and over and under eighteen years.

5. Draft a blank table to present the information below relating to the motor-car industry in the U.K. for the year 1958 and the four years 1968–1971.

Number of workpeople employed.
Average wage earned.
Exports of motor-cars (number and value).
Imports of motor-cars (number and value).
Number of new registrations.

Institute of Company Accountants.

6. Draft a blank table to show the following information for the United Kingdom to cover the years 1939, 1946, 1956 and 1968.

(*a*) Population.
(*b*) Income Tax collected.
(*c*) Tobacco Duties collected.
(*d*) Spirits and Beer Duties collected.
(*e*) Other taxation.

Arrange for suitable columns to show also the "per capita" figures for (*b*), (*c*), (*d*) and (*e*). Suggest a suitable title.

Institute of Company Accountants.

CHAPTER 7

COMPUTATIONS

Comparison of averages. When comparing averages, ratios or percentages, it is very important to be sure that the same things are being compared. If, for example, the average wage of two groups of workers are compared, do the two groups of workers contain the same proportion of men and women? Again, if the rate of profit for a number of years is the subject of comparison, is the composition of the goods sold the same? A change in an average, ratio or percentage may be due to an actual change in the particular thing being compared, or it may be due to a change in the constitution of the groups. When investigating such changes, therefore, it will be necessary to consider to what extent the material is heterogeneous. A good example where care is required is the case of crude death rates. It is often found that the death rate of healthy towns is higher than in slums. The reason is, of course, that in the slums the people are young, whereas in certain very healthy towns such as Worthing the average age is very high.

Another common error is to average a number of averages and to expect that to be the average of the whole. This can easily be shown to give a wrong result.

Example 1.

	Wages	*Employees*	*Average Wage*
Dept. A	£4,000	400	£10
Dept. B	3,200	800	4
Total	£7,200	1,200	£6

£6 which is, of course, the correct average is not the average of £10 and £4.

Mixing of non-comparable records. Suppose the incomes for a certain airline in respect of a certain route, and the gross profits for two consecutive years to be as follows:

Example 2.

	Year 1 £	Year 2 £
Income	140,000	100,000
Gross profit	42,000	24,000
% Profit	30%	24%

Can we say that the rate of profit has fallen?

Consider the following additional analysis:

Example 3.

	Year 1		Year 2	
	Gross profit £	*Income* £	*Gross profit* £	*Income* £
Freight	2,000	20,000	4,000	40,000
Passenger fares	40,000	120,000	20,000	60,000
	42,000	140,000	24,000	100,000

It will be seen that the gross profit on freight is the same for both years, *viz.* 10 per cent; also the gross profit on passenger fares is the same for both years, *viz.* 33⅓ per cent. The ratio of profit to income has altered NOT because of a change in rates of profit but because there is a change in the distribution of sources of income bearing different rates of profit. *No comparison is valid which does not allow for differentiation of sources of income.*

Fallacies about percentages. As an example of an illogical conclusion about percentages, many people would accept the conclusion that, because the percentage of deaths to total injuries for aircraft A is 6·7 per cent, and for aircraft B it is only 2 per cent, aircraft B is the safer machine. However, given the following information, it will be seen that this is not so.

Example 4.

	Miles travelled	*Total injuries*	*Injuries involving death*
Aircraft A	10,000,000	1,500	100
Aircraft B	5,000,000	3,000	60

Aircraft A has only 1 death per 100,000 miles, whereas aircraft B has 1 death for 83,333 miles.

Standardisation. An alternative method to sub-dividing the data until the required degree of uniformity is obtained in order to make valid comparisons, is a process known as standardisation. This gives a single rate or average, instead of rates and averages for each sub-division. The standardised rates are weighted averages of the individual rates, weighted according to a standard distribution. In the case of a standardised death rate, the individual death-rates for each age-group would be weighted by a standard age distribution. In Example 1, when comparing wages over, say, a period of years, so that the average only showed changes in wages, the average wage for each department might be weighted in the ratio of 4 to 8, to give a single average. If the averages of the totals were compared, this would not only show changes in wages, but also show changes in the relative numbers in each department.

Logarithms simplify calculations. If the logarithms of numbers are substituted for actual numbers, the tediousness of multiplication, division, squaring and raising to various powers, extracting of square, cube and other roots is greatly reduced.

Once the method is understood—and it is quite simple—logarithmic calculations are very easy.

What are logarithms? Natural numbers can be stated as powers of 10. Logarithms of the numbers are the exponents of the powers. This will be understood better from examples:

$100 = 10^2$ therefore the logarithm of 100 is 2
$1000 = 10^3$ „ „ 1000 is 3
$10 = 10^1$ „ „ 10 is 1
$1 = 10^0$ „ „ 1 is 0.

When a number is less than 1, the logarithm is negative. Thus for

0·001 the logarithm is -3
0·0001 „ „ -4
0·01 „ „ -2
0·1 „ „ -1.

It will be seen that the logarithms of the above numbers are all integral numbers. The logarithms of all other numbers which are not multiples of 10 are fractions:

For any number between 1 and 10 the log* is between 0 and 1.
For any number between 10 and 100 the log is between 1 and 2, and so on.

* It is usual to write "logarithm of" as log. Thus we say log 100 = 2, and so on.

For any number between 1 and 0·1 the log is between 0 and −1. For any number between 0·1 and 0·01 the log is between −1 and −2, and so on.

Any fractional log is *always* treated as positive, but as a number less than 1 has a negative log, the log has to be written in such a way that the fractional part is kept positive. How this is done will be explained presently.

How to write a logarithm. A logarithm consists of two parts:

(*a*) An integral number like those quoted above, which is called the *characteristic*, which may be positive or negative.

(*b*) A fractional part called the *mantissa*, which is always positive.

To illustrate this, look for the log of 209 in a table of logarithms. We find opposite the number 209·0 the figures ·3201. Now the characteristic of any number is one *less* than the number of digits to the left of the decimal point. 209 has three digits, therefore the characteristic is 2, and we write the log of 209 as 2·3201.

Now examine the number 0·2090. The number is less than 1, therefore the characteristic is negative, and it is one *more* than the number of zeros between the decimal point and the first significant figure of the number. In 0·2090 there is no zero between the decimal point and the first figure (2), hence the characteristic is one more than zero, and is negative, *i.e.* −1, so we write the log thus: $\bar{1}$·3201 (pronounced bar one, point three two nought one). We do not write −1·3201 because only the characteristic is negative.

Take another series of examples:

Log 6436 = 3·8086
Log 643·6 = 2·8086
Log 64·36 = 1·8086
Log 6·436 = 0·8086
Log 0·6436 = $\bar{1}$·8086
Log 0·006436 = $\bar{3}$·8086
Log 0·0006436 = $\bar{4}$·8086, and so on.

How to read logarithmic tables. The tables shown in Appendix A are typical logarithmic tables. More elaborate tables are available, but they have a similar layout.

The first two figures of the number whose logarithm is required are found in the first column. The third figure is found along the top row. The logarithm relating to the first three figures will be found in the row of the first two figures in the column of the third figure. To this logarithm must be added the figure appearing in the same row in the column, belonging to the second group of columns, at the head of which appears the fourth figure of the number whose logarithm is required.

Example 5. Find the logarithm of 3·179. The figure appearing in row 31 under column 7 is 0·5011. In the same row in column 9 of the second group of columns is the figure 12. This is added to 0·5011, giving 0·5023.

How to find a number from its logarithm. Suppose we have the logarithm 2·6321 and we want the number it represents. In the tables we find opposite 6321 the number 4287.

In the logarithm 2·6321 the characteristic (2) is positive, hence the number the log represents must have three digits, so we place our decimal point accordingly, *viz.* 428·7.

If the log had been $\bar{2}$·6321, the number represented by the log must have one zero between the decimal point in the number and the first significant figure; hence we place our decimal point 0·04287; this is the number represented by log $\bar{2}$·6321.

The following further examples will assist:

3·6321 = 4287·0
5·6321 = 428700·0
1·6321 = 42·87
0·6321 = 4·287
$\bar{4}$·6321 = ·0004287
$\bar{1}$·6321 = ·4287
$\bar{3}$·6321 = ·004287

Computations by logarithms. The following rules are all that it is necessary to know:

To multiply numbers. Find their logs and *add* them together. Look in the tables to find the number corresponding to the total, insert the decimal point and you have the required product.

Example 6 (i). Multiply 64·36 × 42·87.
Add the logs of these numbers:

Log 63·36 = 1·8086
Log 42·87 = 1·6321

Log of product = 3·4407

The antilogarithm (*i.e.* the number relating to the logarithm) of 4407 gives the number 2759

∴ product = 2759.

Example 6 (ii). Multiply 6,436 × 0·04287.

Log ·04287 = $\bar{2}$·6321, but for ease of adding or subtracting a logarithm with a negative characteristic we add and subtract 10 to make a positive characteristic, thus:

Log ·04287 = 8·6321 − 10
Log 6436 = 3·8086

Log of product = 12·4407 − 10
= 2·4407
Antilog of 4407 = 2759
∴ product = 275·9.

To divide numbers. Subtract the logarithm of the divisor from the log of the dividend. The difference is the log of the quotient:

Example 7. Divide 161·2 by 37·49.

Log 161·2 = 2·2074
Log 37·49 = 1·5739

Log of quotient = 0·6335
Antilog of 6335 = 4299
∴ quotient = 4·299.

To extract the root of a number. Divide the logarithm of the number by the index of the root; the quotient is the log of the desired root.

Example 8 (i). Find the square root of 183,700.

Log 183700 = 5·2642
Divide by 2 2)5·2642

Log of square root = 2·6321
Antilog of 6321 = 4287
∴ The square root is 428·7.

Example 8 (ii). Find the 5th root of 84,140.

Log 84140 = 4·9250
Divide by 5 5)4·9250

·9850
Antilog of 9850 = 9661
∴ The 5th root is 9·661.

To raise a number to a stated power. Multiply the logarithm of the number by the exponent of the power; the product is the log of the required power.

Example 9. Find the cube of 90.

Log 90 = 1·9542
Multiply by 3 = 5·8626
Antilog of 8626 = 7290
∴ 90 cubed = 729,000.

Hints on using logarithms. When adding, subtracting, dividing or multiplying a logarithm it is essential to remember that the *mantissa is always positive*, even when the characteristic is negative. Thus, the logarithm of ·6436 which is $\bar{1}$·8086 is equivalent to $-1 + 0{\cdot}8086$.

Example 10. Subtract the logarithm of ·6436 from the logarithm of ·04127.

Log ·04127 = $\bar{2}$·6156
Log ·6436 = $\bar{1}$·8086

$\bar{2}$·8070

The subtraction of the mantissa presents no difficulty. In subtracting the last 8 from the 6, it is necessary to add 10 to the 6, and subtract the corresponding 1 from the $\bar{2}$ (that is -2), which then becomes $\bar{3}$ (*i.e.* -3). $\bar{1}$ is now subtracted from $\bar{3}$ (*i.e.* $\bar{2}$, the original figure, less the 1 necessary to compensate for treating ·6 as 1·6 in order to deduct ·8). Deducting $\bar{1}$ from $\bar{3}$ (that is subtracting -1 from -3) gives $\bar{2}$ (*i.e.* -2).

Example 11. Divide $\bar{3}$·3031 by 7.

$$7)\underline{\bar{3}\cdot 3031}$$
$$\bar{1}\cdot 6147$$

This is easy to follow if $\bar{3}$·3031 is rewritten mentally as $\bar{7} + 4\cdot 3031$. (4 is added to ·3031 and taken away from the $\bar{3}$, thus leaving the number unchanged.)

Where only tables of logarithms are provided (and not separate tables of antilogarithms) the number relating to the logarithm, or antilogarithm as it is usually called, is found by reversing the method of finding the logarithm. Thus, in the above example, 6147 will be looked for in the body of the table of logarithms; the nearest figure to this (below) is 6138, found in the column headed 1 in row 41, and the first three figures of the number are therefore 411. The additional 9 required to make 6138 equal to 6147 is found in column 9 of the 2nd group of columns, of row 41 (we must, of course, keep to the same row); the fourth figure of the number required is therefore 9 (the number of the column heading). The four figures of our number are therefore 4119. Since the characteristic is $\bar{1}$, the actual number is ·4119.

Hand electronic calculators. One examination body (the Institute of Chartered Secretaries and Administrators) states on its statistics paper that hand calculating machines may be used. Because this gives the possessor of an electronic calculator a considerable advantage over those that have to rely on tables of logarithms, this practice at the time of writing (1976) is not general. Nevertheless, tables of logarithms, tables of squares and many others are really obsolete; as is also the slide rule, the cost of a really good one being very little less than the less expensive calculators.

Fig. 8 shows the Sinclair Scientific Programmable hand electronic calculator (manufactured by Sinclair Radionics Ltd., Huntingdon, Cambs., England). The keyboard layout is very simple; most keys have two or more functions, enabling a wide range of functions (including programming) to be available from just 19 keys.

Programmability has considerable advantages. With a non-programmable calculator, every step in a calculation demands at least one key-stroke. With a programmable calculator, constants and operations can be stored in the right sequence in the calculator, ready to operate on the variables as they are entered. The task of the operator is reduced to entering the appropriate variables at the appropriate points. Programs may be taken from the program library or devised by the operator. Either way, they are entered simply by keying in a sequence equivalent to the calculation.

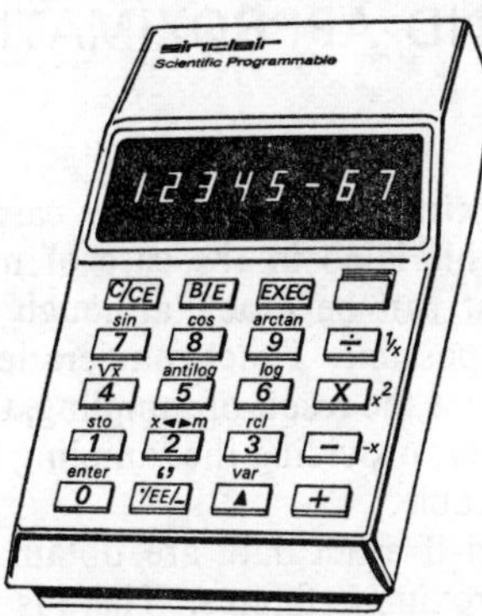

Fig. 8. The Sinclair Scientific Programmable.
Courtesy Sinclair Radionics Ltd.

Questions

1. Multiply 429·9 by 374·9, and 3749 × 0·0429.
2. Divide 78,240 by 67·50, and find the 12th root of 78,240.
3. Cube 485, and divide the result by 25·5.
4. What numbers are represented by the following logarithms?

   3·7596, $\bar{2}$·7596, 0·8129, $\bar{4}$·7930,
   $\bar{1}$·9978, 6·9978, 8·7596, 4·7930.

5. Find the following:

Cube root of 6672.	Square of 599·8.
Square root of 183700.	Cube of 0·05236.
Sixth root of 36720000.	15th root of 8641000.

6. If log 15 is 1·1761, write logarithms of the following numbers.

15,000	1·5	1,500·0
150,000	0·15	0·015
150	0·0015	0·0015

7. A Company has 2 factories. From the data below compare the accident rates in the two factories.

Ages	*Factory X*		*Factory Y*	
	Employees	*Accidents*	*Employees*	*Accidents*
Under 21	200	28	650	91
21–30	250	25	350	35
30–40	500	40	150	12
40 and over	50	2	50	2
	1,000	95	1,200	140

CHAPTER 8

ACCURACY AND APPROXIMATION

The uses of approximation. In many cases, only approximate data are possible. This is so in the case of measurements, for no physical measurement can be exact, although usually the required degree of precision is possible. Exact enumerations are often possible, but again, when they are the result of sampling, they are estimates and approximate. However, providing the sampling is random, the degree of accuracy is measurable.

But even when and if exact data are obtainable, it is often only necessary to use approximate figures. There is no point in spending time and money to get more accurate figures than are necessary for the purpose for which they are required. Often indeed approximate figures enable a clearer picture to be obtained, *e.g.* a firm's Balance Sheet in which the new pence are omitted.

Rounding. When approximate values are given by:

(*a*) *stating the value to the nearest unit*, *e.g.* the nearest ton or the nearest £, or to the nearest multiple of some unit, *e.g.* to the nearest 1,000 tons,

(*b*) *giving the next highest or next lowest whole number*, *e.g.* £3·29 given as £3 or £4,

(*c*) *expressing the value to so many significant figures*, *e.g.* 3·475 expressed correct to 2 significant figures as 3·5,

the figures are said to be rounded.

Note the following apparent error due to rounding:

Example.

Actual	Rounded
3,876	4,000
1,777	2,000
514	1,000
6,167	6,000

The addition of the rounded figures appears to be wrong, but the addition must not be made formally correct, otherwise the total would not be correct to the nearest 1,000.

Statement of degree of accuracy. The extent of the error must be stated. This can be done in a number of ways, *e.g.*:

(*a*) 3,000 tons to the nearest thousand tons,
(*b*) 3,000 ± 500 tons,
(*c*) 3·35 correct to three significant figures,
(*d*) 562 to the nearest whole number,
(*e*) 56 ± 3 per cent.

Kinds of error. The error which arises when an approximate or estimated value is used instead of the actual value must not be confused with errors which arise because of mistakes made in counting, measuring, calculating or in making observations, which are best called mistakes. Error refers to differences between the actual value and the estimated or rounded value.

Such errors may be due to estimating by means of a sample—the smaller the sample, the greater the error; they may arise because of bias or because the data have been recorded approximately.

Relative and absolute errors. The absolute error is the actual difference between the actual value and the estimated or rounded value. The relative error is the ratio, often expressed as a percentage, of the absolute error to the actual figure or where, as is often the case, this is not known, to the estimated value:

Output is 40,000 tons to the nearest 1,000 tons,
Maximum absolute error is 500 tons,
Maximum relative error is 500 divided by 40,000, *i.e.* 0·0125 or 1·25 per cent.

Laws of errors.

1. *The absolute error of a sum equals the sum of the absolute errors of its components.*

Add 500 (to the nearest 10) and 400 (to the nearest 100).

$$(500 \pm 5) + (400 \pm 50) = 900 \pm 55.$$

2. *The absolute error of a difference equals the algebraic difference of the errors of its components.*

From 500 (to the nearest 10) subtract 400 (to the nearest 100).

$$(500 \pm 5) - (400 \pm 50) = 100 \pm 55.$$

3. *The relative error of a product is equal to the sum of the relative errors of its components.*

Multiply 500 (to the nearest 10) by 400 (to the nearest 10).

$$(500 \pm 1\%) \times (400 \pm 1\tfrac{1}{4}\%) = 200{,}000 \pm 2\tfrac{1}{4}\%.$$

4. *The relative error of a quotient is equal to the algebraic difference of the relative errors of its components.*

Divide 500 (to the nearest 10) by 400 (to nearest 10).

$$(500 \pm 1\%) \div (400 \pm 1\tfrac{1}{4}\%) = 1{\cdot}25 \pm 2\tfrac{1}{4}\%.$$

The last two laws are only approximately true, and then only provided that the degree of error is small.

Note that in the example given under rule 4 it is the possible error that must be calculated, *i.e.* the greatest error possible. This will be when 500 + 1 per cent is divided by 400 — $1\frac{1}{4}$ per cent; the relative error is then 1 per cent — (—$1\frac{1}{4}$ per cent), *i.e.* $2\frac{1}{4}$ per cent. Similarly, the relative error of the other limit is —1 per cent — (+$1\frac{1}{4}$ per cent), *i.e.* —$2\frac{1}{4}$ per cent.

Biased error. When the error is all in one direction, it is said to be biased or cumulative. Thus amounts which are given to the next highest whole number are biased.

Compensating errors. When the errors are such that they tend to cancel each other, they are said to be unbiased or compensating. Thus, numbers approximated to the nearest whole number are likely to give rise to compensating errors.

Accuracy and errors. When errors are biased, the absolute error of a sum will be greater than the errors of the individual items—the greater the number of items, the greater the error. In the case of subtraction, the absolute error may be reduced. In the case of multiplication, the relative error will be increased, but in the case of division, the relative error will be decreased.

Where the number of quantities is large and the errors are compensating, the resultant error of their sum or difference is reduced.

Questions

1. Explain the meaning of the statistical terms:
   (*a*) compensating error, and
   (*b*) biased error.

The numbers 56,000, 7,000 and 20,000 are correct to 5 per cent, 0·5 per cent and 0·05 per cent respectively. Calculate the absolute error in the aggregate of the three numbers. Express this absolute error as a percentage of the aggregate value.

2. Explain "statistical error." Approximate the following figures to:
   (*a*) Nearest thousand £.
   (*b*) Nearest hundred £.
   (*c*) Next thousand £ below.

Calculate the absolute and relative errors of their totals.

Exports of Mining Machinery from the United Kingdom, May 1972

To West Africa	34,053
„ South Africa	12,865
„ India	33,233
„ Malaya	8,177
„ Australia	45,847
„ Other Commonwealth countries and Irish Republic	67,581
„ Poland	12,461
„ France	18,346
„ Other foreign countries	78,455

Institute of Company Accountants.

3. Give specimen illustrations of the meaning of the term "error" in statistical work and, hence, show how this differs from the ordinary interpretation of the word.
Chartered Institute of Transport.

4. Define statistical error. Find the values of:

$(19 \pm 2) + (85 \pm 6)$
$(76 \pm 5) - (15 \pm 1)$
$(40 \pm 7) \times (25 \pm 4)$
$(480 \pm 20) \div (120 \pm 5)$

Institute of Company Accountants.

5. In statistical work it is often necessary to use approximate figures. Show how the degree of accuracy in the result can be determined in the course of arithmetical manipulation of such data.
Local Government Examination Board.

6. What do you understand by the estimation of errors?
The total profits of 144 companies amount to £7,425,000. Estimate the possible error when each company's profits have been approximated to:

(i) the nearest thousand £;
(ii) the next thousand £ above the actual figure.

Institute of Company Accountants.

7. Rewrite the following table giving the figures correct to the nearest 1,000. What is the amount of:

(*a*) the absolute error of the total, and
(*b*) its relative error?

Suppose *your* figures were the only ones available; then state:

(*c*) the percentage of emigrants who went to Canada, and
(*d*) the amount of relative error in the above percentage, and show its relation to the relative error of the number who went to Canada and the relative error of the total.

Emigrants of British Nationality Travelling Direct by Sea from the United Kingdom to Extra-European Countries, 1950

By Destination

Canada	13,434
Australia	54,184
New Zealand	10,562
South Africa	9,320
Other British countries	25,434
Foreign countries	17,304
	130,238

Source: *External Migration, H.M.S.O.*
Institute of Statisticians.

CHAPTER 9

DESIGN OF FORMS

Much of the information that is subsequently summarised and presented in the form of tables is first recorded on a form. One type

TYPE OF ENTRY SPACES

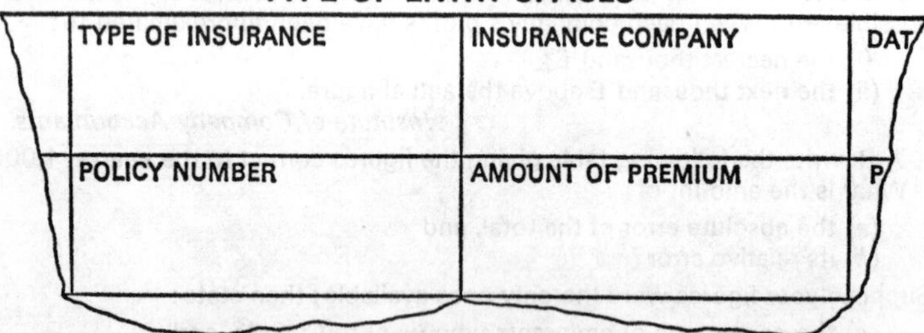

TYPE OF INSURANCE	INSURANCE COMPANY	DAT
POLICY NUMBER	AMOUNT OF PREMIUM	P

Fig. 9.—Panel arrangement.

(a) NAME (b) ADDRESS	(a) JOB (b) DEPARTMENT
(a)	(a)
(b)	(b)
(a)	(a)
(b)	(b)
(a)	(a)

Fig.10.—Columnar entries.

Notes: (1) Each column is used for two entries, thus avoiding too wide a form.
(2) The shading in the first column makes the name stand out more easily in a long list.

of form is the questionnaire. But in any case, whatever the type of form, there are certain principles to be observed in its design.

The purpose of forms. In a Stationery Office publication, *The Design of Forms*, it is stated, "Forms are used to record and communicate information in order that the information shall be set out in an orderly, pre-determined way and a regular routine established for entering and using the information and handling the form itself."

How to design a form. Before the actual designing of the layout, careful consideration must be given to the following matters, all of which affect the design if the form is to fulfil its proper functions.

(*a*) The information wanted.
(*b*) How will the form be used? Who will enter the information, method of sorting, tabulating, filing, where the form is to be used, *e.g.* in the factory, outdoors, the frequency of use.

MARK SHEET													
NAME OF STUDENT	MATHS	ENGLISH	HISTORY	GEOGRAPHY	CHEMISTRY	PHYSICS	ECONOMICS	FRENCH	MUSIC	P.T.	DRAWING	SCRIPTURE	LA

Fig. 11.—Headings for narrow columns.

In the actual designing of the layout, careful attention should be paid to the following points:

(*a*) The form should not look complicated.
(*b*) The captions should be clear and precise.
(*c*) Instructions (which should be adequate) should, if lengthy, be given in a part of the form distinct from the questions.
(*d*) Give the form some identification (*e.g.* a title).
(*e*) Arrange the entries so that the form is easy to read and it is easy to make the entries.
(*f*) Spacing must receive special care. Remember this is determined by the entry to be made, not by the length of the caption.

TYPE OF DWELLING

FLAT ☐	TERRACED HOUSE ☐	DETACHED ☐
BUNGALOW ☐	SEMI-DETACHED ☐	

Fig. 12.—"Boxes." The appropriate box will be "ticked."

Details	Last month		This month		Increase or Decrease	
	£	%	£	%	£	%
Gross sales						
Less returns and allowances						
Net sales						
Less prime cost of sales						
Gross profit on sales						
Less selling expenses						
Less general expenses						
Net profit on sales						
Add other income						
Net income						
Less fixed charges						
Add balance brought forward						
Amount available						
Appropriations						
To carry forward						

Fig. 13.—An example of a form used by accountants.

Questions

1. Design a form for the collection, from a number of provincial distributive branches, of information which is to be used as a guide to production policy. on sales of clothing. *Institute of Statisticians.*

2. A manufacturing organisation has selling branches in each large town in the country. It makes 6 kinds of articles which are sold retail and wholesale by the branches. The Head Office wishes to plan a Sales Campaign based on the past sales and likely future demand. Design a form for the collection of the necessary data and draft the instructions for completing the form. *Institute of Statisticians.*

3. Design a form for recording the daily output from a set of six similar machines, with spaces for entering monthly totals and appropriate comparisons of performance of the different machines. *Institute of Statisticians.*

CHAPTER 10

CHARTS AND DIAGRAMS

Statistical data can often be presented in chart form or by means of diagrams or graphs, in addition to being presented in tabular form. Often this enables relationships and trends and comparisons to be grasped more readily.

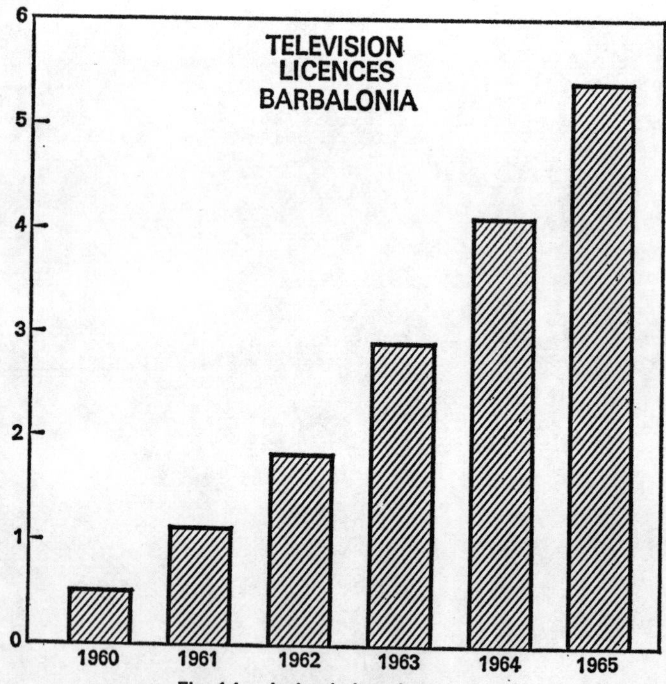

Fig. 14.—A simple bar chart.

Bar charts. These charts enable magnitudes to be compared visually. Bars are drawn whose length is proportional to the magnitude to be represented. Thus, if it were required to compare a sales figure of £2,000 with a sales figure of £2,500, two bars would be drawn whose lengths were proportional to these two amounts, for example, 2 inches and $2\frac{1}{2}$ inches respectively.

When data are to be presented in the form of bar charts, a suitable scale must be chosen, and this will be indicated either at the side or the bottom of the diagram. This scale MUST start at zero, otherwise false impressions are given. The choice of horizontal or vertical scales is optional when items in a group are to be compared, as, for example, a bar chart comparing the populations of the continents. When, however, the data are to be charted with reference to a series of dates, the bars should be drawn vertically, so that the dates appear horizontally, and the scale is shown vertically (see Fig. 14).

The bars are usually separated, the width is entirely a matter of neatness of presentation. A chart, like a table, must have a title.

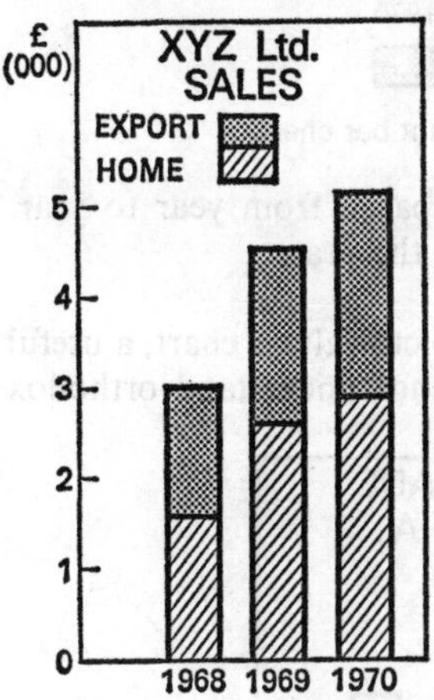

XYZ Ltd.—Sales (£000)

	1968	1969	1970
Export	1·4	1·8	2·28
Home	1·6	2·7	2·9
Total	3·0	4·5	5·18

Note the key and the method of shading.

Fig. 15.—A component bar chart.

Component bar charts. These charts show visually the way in which a whole picture is divided. Thus a bar chart, representing sales, might be divided into two parts, representing home sales and export sales. Of course, the divisions would be proportional to the respective sales.

The percentage component bar chart. In this case, the bars are divided in proportion to the percentages that the parts bear to the whole. The scale will be a percentage scale, and all charts will be the same length. A key will also be necessary (Fig. 16).

The multiple bar chart. This is sometimes known as a compound bar chart. This chart groups two or more bar charts together. More than one set of comparisons can be made. Fig. 17 gives an example of such a chart. Here exports and imports can be compared from year

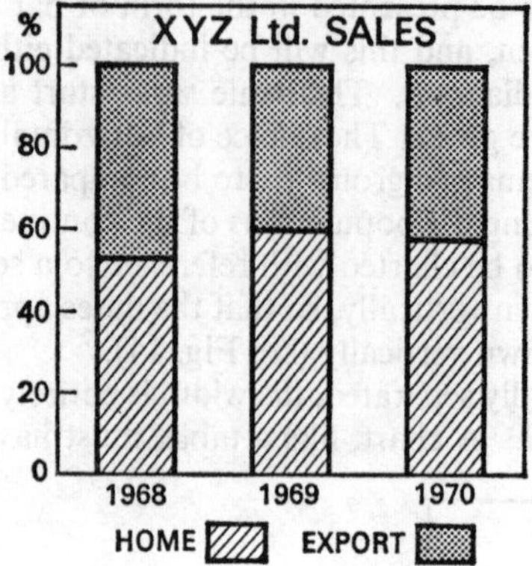

Fig. 16.—A percentage component bar chart.

to year. In addition, imports can be compared from year to year. Finally, exports can also be compared over the years.

Pictograms. These are really a form of pictorial bar chart, a useful way of presenting data to people who cannot understand orthodox

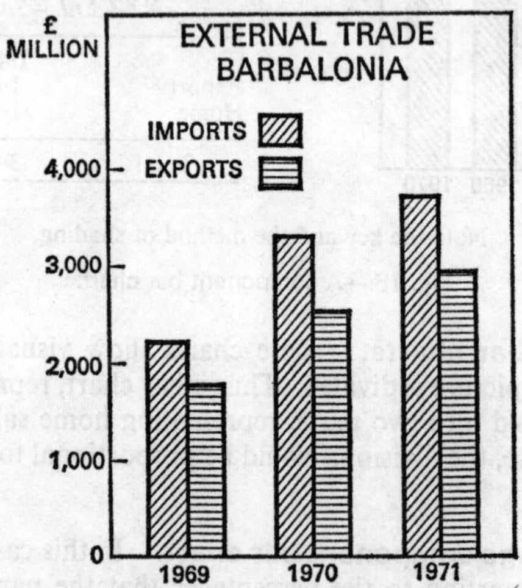

Fig. 17.—A multiple bar chart.

charts. Small symbols or simplified pictures represent the data. There are important rules to follow if pictorial charts are to fulfil their function.

(*a*) The symbols must be simple and clear.
(*b*) The quantity each symbol represents should be given.

(c) Larger quantities are shown by a greater number of symbols, and not by larger symbols. This is a frequent error. A part of a symbol can be used to represent a quantity smaller than the value of a whole symbol.

Pictograms are used to make visual comparisons in exactly the same way as a bar chart, and, in effect, the number of symbols corresponds to the length of a bar chart (Fig. 18).

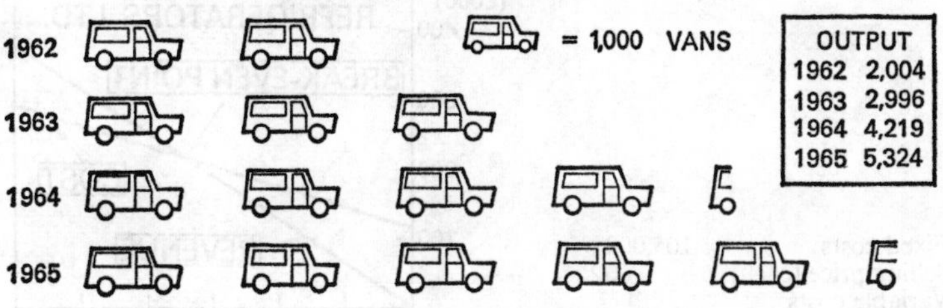

Fig. 18.—A pictogram.

Pie charts. Like the component bar charts, pie charts show the relationship of parts to the whole. However, there is one important difference. In the case of component bar charts, the length of bars are

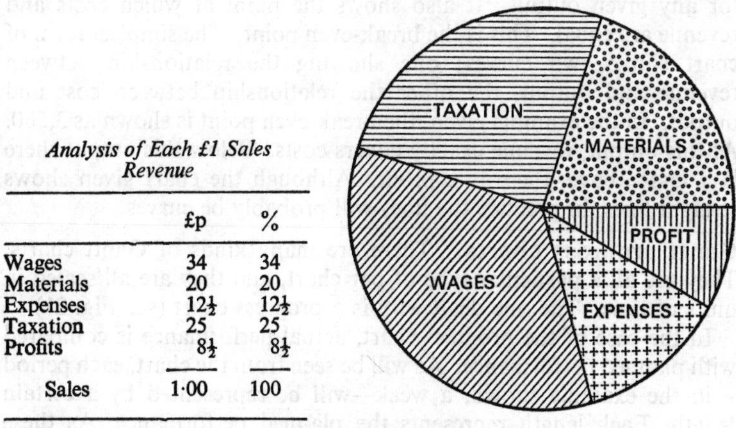

Analysis of Each £1 Sales Revenue

	£p	%
Wages	34	34
Materials	20	20
Expenses	12½	12½
Taxation	25	25
Profits	8½	8½
Sales	1·00	100

Fig. 19.—A pie chart.

compared, whereas in the case of pie charts, areas of segments are compared. It is, however, difficult to compare areas visually. For this reason pie charts are an inferior form of presentation. They are extremely popular, except among statisticians. Fig. 19 is an example of a pie chart. Often the percentages are inserted in the segments, so that, in effect, it is the figures that are compared.

Construction of a pie chart. A circle consists of 360°. The proportion that each part bears to the whole will be the corresponding proportion of 360°, which will require to be calculated. In the example given, taxation is one-fourth of the whole. The segment relating to taxation will therefore have an angle of one-fourth of 360°, *i.e.* 90° at the centre.

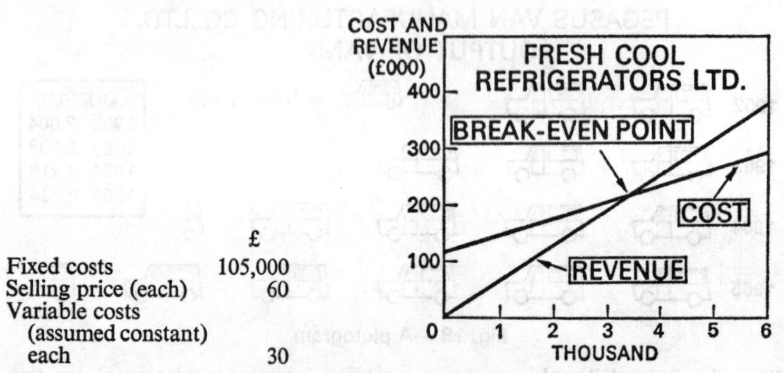

Fig. 20.—A break-even chart.

Break-even charts. A break-even chart shows the profit or loss for any given output. It also shows the point at which costs and revenue are equal. This is the break-even point. The simplest form of chart shows two curves; one showing the relationship between revenue and output, the other the relationship between cost and output. In the example given, the break-even point is shown as 3,500. At this output, revenue exactly covers costs. Below this output there is a loss; above, there is a profit. Although the chart given shows straight lines, in practice the lines will probably be curves.

Gantt progress charts. There are many kinds of Gantt charts. They are all a particular type of bar chart, and they are all scaled in units of time. The one dealt with is a progress chart (see Fig. 21).

In the case of the progress chart, actual performance is compared with planned performance. As will be seen from the chart, each period —in the example shown, a week—will be represented by a certain length. Each length represents the planned performance. As these are not necessarily equal, the same length, representing the same length of time, does not represent the same quantities. The actual performance is charted, and in addition, the cumulative performance is shown. The weekly results are shown by a thin line, the cumulative by a thick line (see Fig. 21).

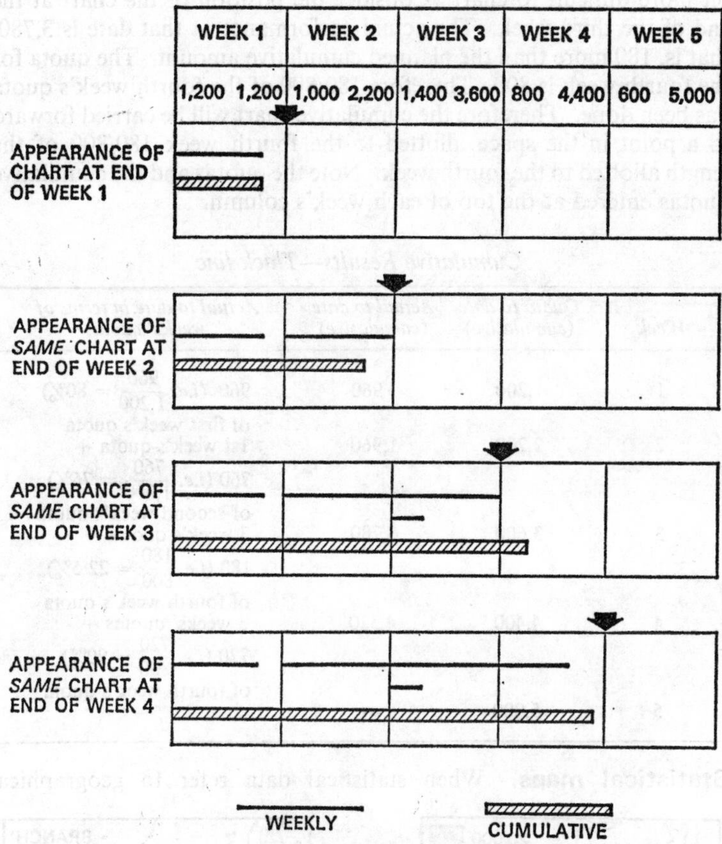

Fig. 21.—A Gantt progress chart.

Output
Weekly Results—Thin Line

Week	*Quota*	*Actual*	*Actual as % of quota*
1	1,200	960	80
2	1,000	1,000	100
3	1,400	1,820	130
4	800	540	67·5
5	600	—	

In the first week only 80 per cent of the planned performance was achieved; hence the line charted to show the actual performance is 80 per cent of the length allocated to week 1. In the case of week 3, more than one line is required. The cumulative line, the thick one, is a bit more difficult to chart. Consider the position of the chart at the end of the third week. The actual performance at that date is 3,780, that is, 180 more than the planned cumulative amount. The quota for the fourth week is 800. Therefore 180/800 of the fourth week's quota has been done. Therefore the cumulative chart will be carried forward to a point in the space allotted to the fourth week 180/800 of the length allotted to the fourth week. Note the quotas and the cumulative quotas entered at the top of each week's column.

Cumulative Results—Thick line

Week	*Quota to date (cumulative)*	*Actual to date (cumulative)*	*Actual to date in terms of weeks' quotas*
1	1,200	960	960 (*i.e.* $\frac{960}{1,200}$ = 80%) of first week's quota
2	2,200	1,960	1st week's quota + 760 (*i.e.* $\frac{760}{1,000}$ = 76%) of second week's quota
3	3,600	3,780	3 week's quotas + 180 (*i.e.* $\frac{180}{800}$ = 22·5%) of fourth week's quota
4	4,400	4,320	3 weeks' quotas + 720 (*i.e.* $\frac{720}{800}$ = 90%) of fourth week's quota
5	5,000	—	—

Statistical maps. When statistical data refer to geographical

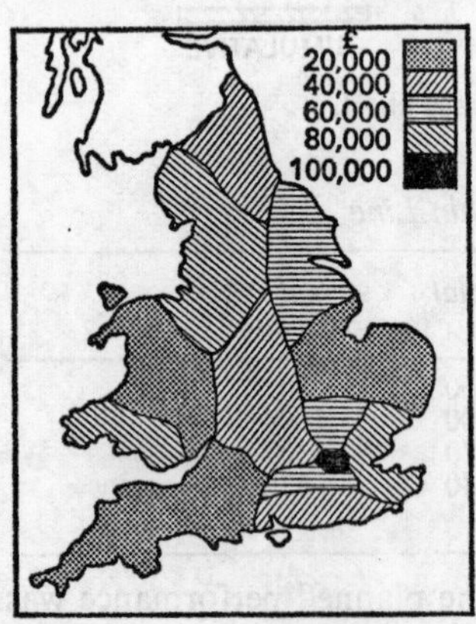

Fig. 22.—Sales according to salesmen's area.

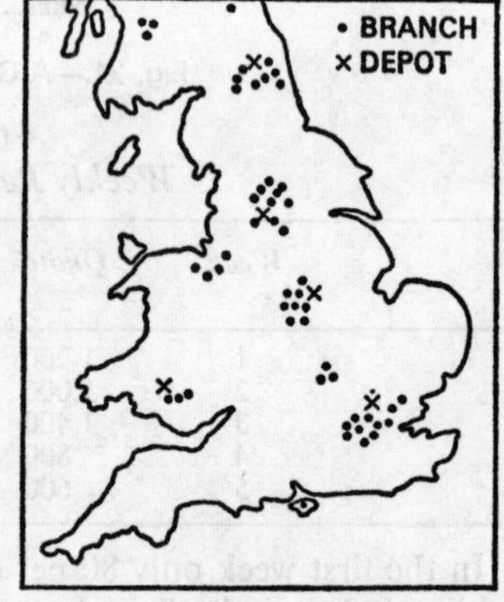

Fig. 23.—Distribution of branches.

areas, often a suitable form of presentation is the statistical map or cartogram.

The main types of such maps are:

1. *The hatched or shaded map.* The varying size of the data is denoted by different shading or hatching—the greater the size, the denser the shading or the closer the hatching. Fig. 22 shows such a map. This type of map is also used to show averages and ratios relating to various areas.

2. *Dot maps.* These show the number of occurrences. They may be placed at their exact location, as in Fig. 23, or by a single dot placed in the area to which they refer, varying in size according to the number they represent, as in Fig. 24.

3. *Maps presenting graphs or charts.* In many cases, charts or graphs can be placed on the areas to which they refer. Fig. 25 is an example of such a map.

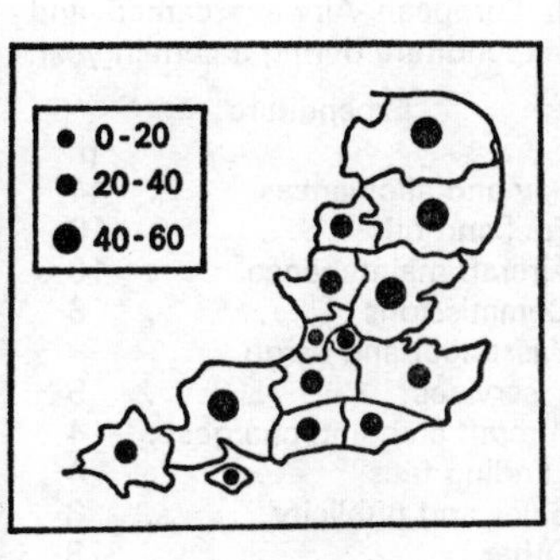

Fig. 24.—Number of salesmen in south-eastern counties.

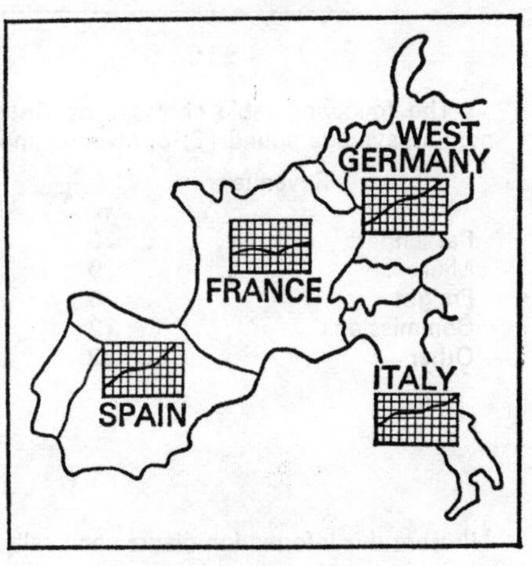

Fig. 25.—Growth of sales to certain European countries.

Questions

1. *Exports of Textiles*, 1971
(*Woven Piece Goods—Million Square Yards*)

	April	*May*	*June*	*July*	*Aug*	*Sept*	*Oct*	*Nov*	*Dec*
Cotton	96	78	72	65	77	71	67	73	53
Wool	13	10	9	10	10	9	8	7	6

Make a graphical comparison, by means of a compound bar chart, of the volume of exports given in the above table. Use the chart to write a brief report on the exports of woven piece goods during the last nine months of 1971.

2. Show the following information on a suitable diagram and give comments upon any features of interest to the transport industry:

Estimated Population in Zones (000)

Zone radius (*miles*)	*Manchester* 1950	*Manchester* 1975	*Stockholm* 1950	*Stockholm* 1975
1	231	199	28	15
2	415	274	212	190
3	342	319	162	172
4	265	270	87	96
5	80	82	73	97
6	32	36	86	123
7	28	32	72	115
8	3	4	63	135

Chartered Institute of Transport.

3. The following table shows how British European Airways earned and spent an average pound (£) of revenue and expenditure during a certain year.

Revenue	p	Expenditure	p
Passenger	79	Pay and allowances	44
Mail	9	Fuel and oil	16
Freight	8	Aircraft maintenance	10
Commissions	2	Commissions	6
Other	2	Passenger and cargo services	5
		Aircraft standing charges	4
		Landing fees	4
		Sales and publicity	3
		Other	8

Present this information diagrammatically:

(*a*) using Pie-diagrams, and
(*b*) using any other form of diagrammatic presentation of your own choice, giving reasons for your choice. *Chartered Institute of Transport.*

4. Describe two forms of percentage distribution diagrams.

5. Describe each of the diagrams listed below and give an illustration in each case:

Bar diagram	Break-even chart
Compound bar diagram	Percentage bar diagram
Progress chart	Pictogram

6. How would you present statistical data referring to geographical areas on maps?

7. **The budgeted production and the actual quantities produced of a certain component manufactured by XL Motors Ltd. were as follows:**

Month	*Budget*	*Actual*
January	1,000	800
February	1,500	1,600
March	2,000	1,800
April	2,000	2,100
May	2,500	2,800
June	3,000	2,900

Show the Gantt progress chart as it would appear at the end of (*a*) April, (*b*) June.

CHAPTER 11

NETWORK PLANNING

Any operation, project or task which is undertaken—however simple or complicated—can be divided into a number of activities. Building a house, for example, could be analysed into a large number of activities, among which would be found, say, drawing up the plans, digging the foundations, ordering bricks and so on. Some activities must be completed before others are started; some activities can be carried on at the same time; activities must be performed in an orderly manner so that the project can be completed in the most efficient way.

When the project consists of a large number of activities planning becomes extremely difficult. *Network analysis* provides a technique for efficient planning of any operation. It can also be used very effectively for comparatively small tasks which are usually done by rule-of-thumb methods.

Data required to draw a network. In respect of *each* activity it is necessary to know:

(*a*) *The immediately preceding activity* (*or activities*). As soon as the immediately preceding activity (or activities) are finished it is possible for the activity in question to be started.

(*b*) *The immediately succeeding activity* (*or activities*). It follows that if activity E immediately precedes C then C immediately follows E (see Fig. 26).

PROJECT ANSCO

Activity	*Immediately preceding activities*	*Immediately following activities*	*Duration of activity (days)*
A	C	D H	3
B	E	F	7
C	E	A	12
D	A	F	5
E	G	B C	8
F	B D H	None	4
G	none	E	2
H	A	F	9

Fig. 26.—Data required to draw a network.

(*c*) *The duration of the activity.* This is the time required to complete the activity.

Symbols used in drawing a network. An activity is denoted by an arrowed line joining two circles known as *nodes* or *events* which denote the start and finish of the activity. The events are numbered; an activity can thus be identified by the numbers in the nodes at the start and finish of the activity, *e.g.* 5–6. This is shown in Fig. 27.

Fig. 27.—Symbols used to denote an activity.

The arrow points towards the immediately following activity; it must never point towards the preceding one.

The use of dummy activities in a network. A *dummy activity* represented by a broken arrowed line is a non-existent activity, and thus takes no time to perform and uses no resources.

An activity is identified by the numbering of the nodes at the start and finish of the activity.

Where activities can be carried out simultaneously, that is, each activity has the same preceding activities and the same following activities, then, since each activity must have a different identity reference, a dummy activity is introduced into the network.

In project Ansco (see Fig. 26), where activities D and H are both preceded by A and both followed by F, a dummy activity is introduced into the network as shown in Fig. 28.

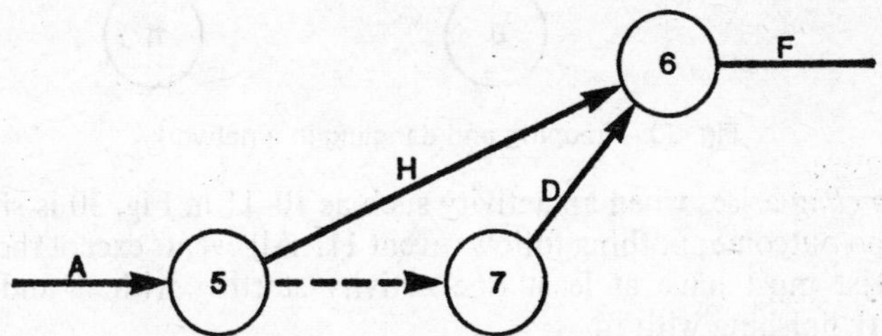

Fig. 28.—Dummy activity to ensure activity identity.

Consider an activity A which immediately follows activities X *and* Y and an activity B which immediately follows *only* X. Again it is necessary to introduce a dummy activity—this time to ensure that the network reflects the facts. This will be clear from Fig. 29.

The direction in which the dummy arrow points is very important.

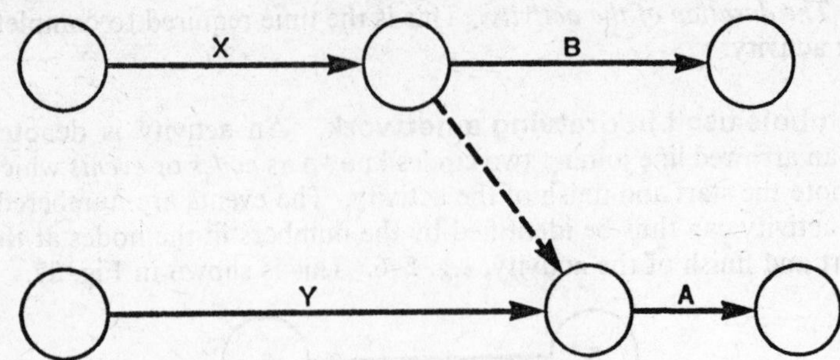

Fig. 29.—Dummy activity to preserve logicality.

There is a tendency for students to introduce unnecessary dummies. They must not be introduced other than for the two purposes just described.

Looping and dangling. These errors must be avoided in drawing the network. *Loops* are illogical. Referring to Fig. 30, event 9 occurs after event 7 and event 8 after event 9; event 8 must therefore occur after event 7 and *not* before as shown in the network.

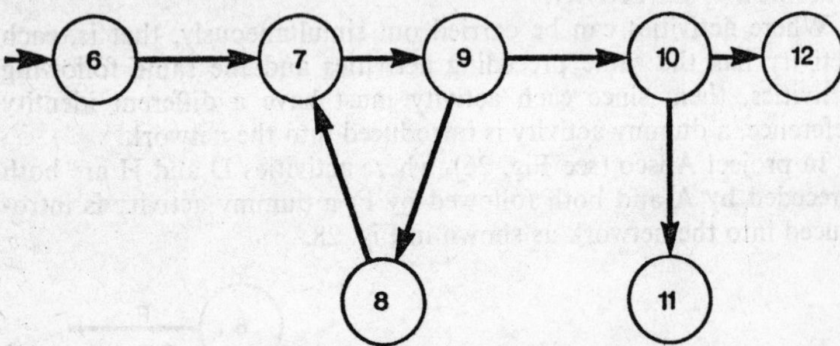

Fig. 30.—Looping and dangling in a network.

Dangling arises when an activity such as 10–11 in Fig. 30 is shown with no outcome; nothing follows event 11. All events except the first and last must have at least one activity starting with it and one activity finishing with it.

A typical network. Fig. 31 shows the network drawn from the data given in Fig. 26. The events have been numbered from left to right and from top to bottom of the network. This is the usual method of numbering.

The activity has been described—in this case by a letter—above the arrowed line and below the line the duration of the activity has been

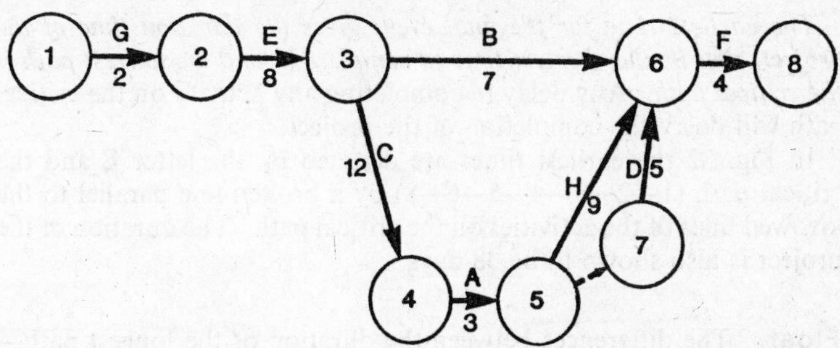

Fig. 31.—Network for Ansco project.

given. *Note that the length of the arrowed lines is NOT proportional to the duration of the activity.* The lines are always drawn straight and not curved.

The critical path and the duration of project. Starting from event 1 and calling this zero time, then the earliest time event 2 can take place, that is the earliest time activity E can start is $0 + 2 = 2$ days after event 1 when activity G was started, since the duration of G is 2 days. The earliest time event 3 can take place, that is, the earliest time B and C can start is $2 + 8 = 10$ days after event 1. The earliest time event 6 can take place if only activity B precedes it is $10 + 7 = 17$ days, *but* this is not the case. C, A and H also precede it and all these activities have to be completed before F can start. The earliest time event 6 can take place is therefore $10 + 12 + 3 + 9 = 34$ days after event 1.

Where there are two or more paths to an event the earliest time for the event is the time taken by the longest path.

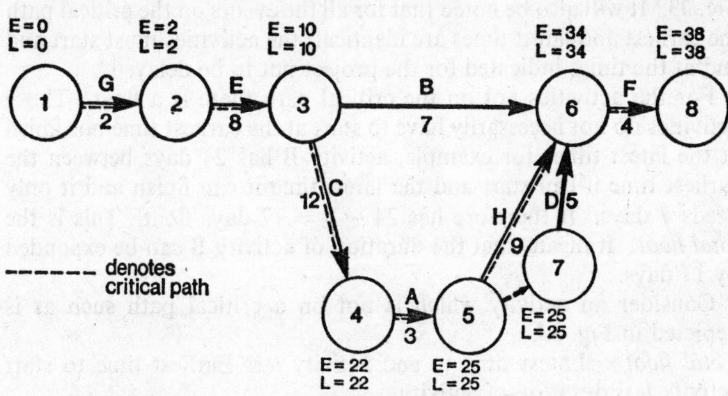

Fig. 32.—The critical path, Ansco project.

The earliest time for the final event gives the duration time of the project, that is, the shortest time to complete it, and the longest path is the critical path. Any delay in completing any activity on the critical path will delay the completion of the project.

In Fig. 32 the earliest times are denoted by the letter E and the critical path (1—2—3—4—5—6—8) by a broken line parallel to the arrowed lines of the activities on the critical path. The duration of the project is also shown to be 38 days.

Float. The differences between the duration of the longest path—the critical path—and the alternative paths are slacks or floats; the activities on these alternative paths except those shared with the longest path can be delayed without affecting the duration of the project.

Calculating the amount of float in respect of all those activities not on the critical path enables the planner to know the extent of the excess resources, men and/or materials, and with what activities they are associated.

Calculation of float. The earliest time to finish the project is also the latest time to finish it. In project Ansco this is 38 days. The latest time for event 6 to occur if the project is to finish on time is $38 - 4 = 34$ days since activity takes 4 days. The latest time event 3 can take place if only activity B follows it is $34 - 7 = 27$ days. But C, A and H also follow. The latest time for event 3 is therefore $34 - 9 - 3 - 12 = 10$ days. *Where there are alternative paths from an event, the latest time for an event is the lowest figure.* It will be noted that the latest times for an event to take place, that is for the following activity or activities to start, is denoted by L on Fig. 32 and Fig. 33. It will also be noted that for all the events on the critical path the earliest and latest times are identical; the activities must start and end at the times indicated for the project not to be delayed.

For the activities not on the critical path there is a float. Those activities do not necessarily have to start at the earliest time nor finish at the latest time; for example, activity B has 24 days between the earliest time it can start and the latest time it can finish and it only needs 7 days. It therefore has $24 - 7 = 17$ days float. This is the *total float.* It means that the duration of activity B can be expanded by 17 days.

Consider an activity which is not on a critical path such as is depicted in Fig. 33.

Total float = Latest time to end activity *less* Earliest time to start activity *less* duration of activity.

In the case of the activity depicted in Fig. 33 this is $40 - 8 - 15 = 17$.

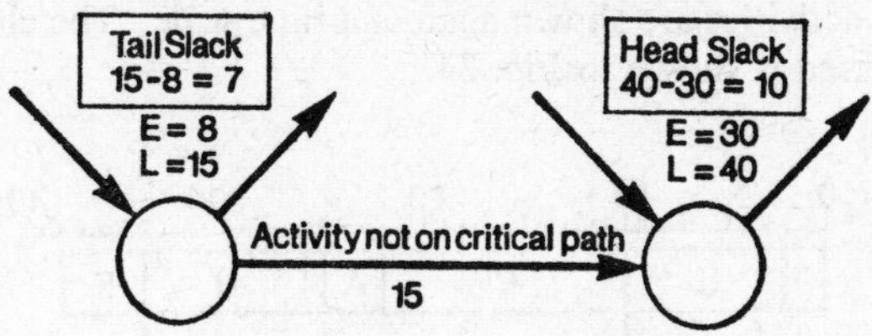

Fig. 33.—An activity with float.

This is the total amount of time by which the activity could be expanded without affecting the duration of the project, *but* only if the activity started at the earliest time and ended at the latest time. This condition could well affect the floats of the other activities, that is the time by which they could be expanded.

Free float is the amount the activity can expand without affecting the float of the following activity.

Free float = Total float *less* Head slack (that is the latest time an activity can finish *less* the earliest).

In this case, the head slack is 40 − 30 = 10 and the free float is therefore 17 − 10 = 7.

Independent float is the amount an activity can expand without affecting the floats of *both* the previous and following activities.

Independent float = Free float *less* Tail slack (that is the latest time an activity can start *less* the earliest).

In this case, the tail slack is 15 − 8 = 7 and the independent float is therefore 7 − 7 = 0.

Listed below are the activities of project Ansco which are *not* on the critical path together with their earliest and latest starting and finishing times and also their duration (see Fig. 32).

	Starting time		*Finishing time*		
Activity	*Earliest*	*Latest*	*Earliest*	*Latest*	*Duration*
B	10	10	34	34	7
D	25	25	34	34	5

The *free float* is calculated as follows:

Activity	*Total float*	*Head slack*	*Free float*
B	34 − 10 − 7 = 17	34 − 34 = 0	17 − 0 = 17
D	34 − 25 − 5 = 4	34 − 34 = 0	4 − 0 = 4

Sequenced Gantt chart. A very useful way of presenting the planning and scheduling of a project is by means of a Gantt bar chart where the activities are shown against a time scale. The chart for the project Ansco is shown in Fig. 34.

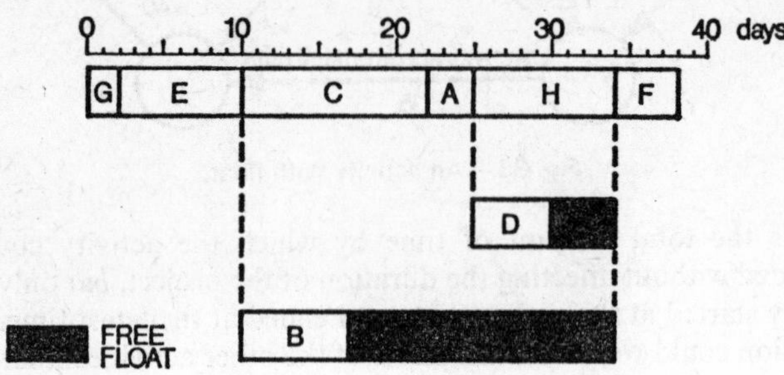

Fig. 34.—Sequenced Gantt chart.

The activities comprising the critical path constitute the first bar. The other activities will be represented by other bars parallel to the first bar and placed beneath it; each bar representing a branch of the network. The length of the bars are proportional to the duration of the activities *plus* their free float. The relationships between the branches are indicated by vertical broken lines.

Questions

1.

The "AJAX" project.

Activity	Immediately preceding activities	Immediately succeeding activities	Duration of activity days
A	none	B, D	3
B	A	C, G	6
C	B	I	8
D	A	E, F	2
E	D	H	5
F	D	H	4
G	B	H	2
H	E, F, G	none	9
I	C	none	12

You are required to: (*a*) draw the network; (*b*) Indicate the critical path; (*c*) draw a sequenced Gantt chart.

2. A company has decided to build new offices on a plot of land which it owns, and a list of the necessary activities has been drawn up by the building contractor. These activities are given in the table below:

Activity	*Description*	*Preceding activities*	*Duration (weeks)*
I	Draw up plans	—	3
II	Order furniture and office equipment	I	1
III	Level the site	I	3
IV	Mark out site	III	2
V	Lay drainage	IV and II	4
VI	Make approach roads	IV and II	3
VII	Lay foundations	V	7
VIII	Erect walls	VII	12
IX	Lay paths	VI and VIII	4
X	Erect roof	IX	2
XI	Erect internal walls	X	5
XII	Complete building	XI	10

Set out the network for these activities and identify the critical path. Thus indicate the least time which will be taken to complete the new offices on the basis of the above durations.

I.C.S.A. Part 1, Dec. 1975.

3. Following the receipt of an application for a mortgage loan by an insurance company, it is found that one day is taken in establishing all the facts required to identify the nature of the property and the amount of the applicant's income. Credit references are then taken up, and at the same time surveyors are asked to value the property. It takes 5 days for references to be obtained and 7 days for a survey to be completed. When credit references and a survey are available the application can proceed to the mortgage committee, and it takes 5 days for a decision to be made. Whilst committee consideration is being given the Accounting Dept. is being consulted to determine when funds will be available for the loan, and the rate of interest to be charged. It takes three days for the Accounting Dept. to make its decision. Finally, when committee approval has been gained, and it is known that funds are available, the application can proceed to the Board of Directors for sanction. It takes 7 days for this last process to be complete. How long does it take from the receipt of the application form to final agreement by the Board to gain sanction for a loan? Show the details of your estimate in the form of a network diagram and indicate the critical path.

I.C.S.A. Part 1, June 1975.

CHAPTER 12

GRAPHS

The movement of data can be presented very effectively by means of a graph, sometimes known as a line chart.

The purpose of a graph. A graph shows, by means of a curve or straight line, the relationship between two variables; for example, the amount of sales and the period of time when the sales were made, or

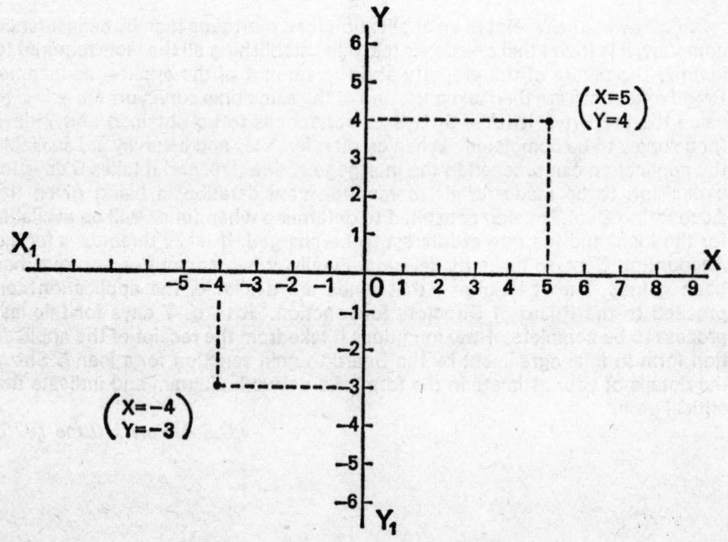

Fig. 35.—The construction of graphs.

the relationship between output and cost. One of the variables will be the independent, and the other the dependent, variable. The formulation of the problem will determine which variable is chosen as dependent. For example, the relationship between the number of bushels of wheat per acre and the amount of rainfall can be formulated in two ways. Given the rainfall, what is the output per acre? or, given the output per acre, what is the rainfall? The first would be the more usual way. The output per acre would depend on the rainfall. The output would be the dependent variable, the rainfall, the independent variable.

Historigrams. When one of the variables changes over time, *e.g.* population, or sales, *i.e.* when the values of the variable form a times series, the graph which shows this relationship is known as a historigram (see Figs. 36 and 42).

(*Note:* Do not confuse a historigram with a histogram, an entirely different type of diagram, illustrated in Fig. 51.)

Construction of graphs. The first step in the construction of a graph is the drawing of two lines at right angles. These lines are known as co-ordinate axes (see Fig. 35).

The axis YOY_1 is called the ordinate or, less mathematically, the vertical axis; XOX_1 is called the abscissa, or the horizontal axis. Alternate appellations are the y-axis and x-axis respectively.

Both these axes will have a scale, chosen so that the graph will fit the paper, will include all the observations and will avoid too sharp or too flat a curve.

Daily Output

Day	*Output*
1	100
2	300
3	600
4	500

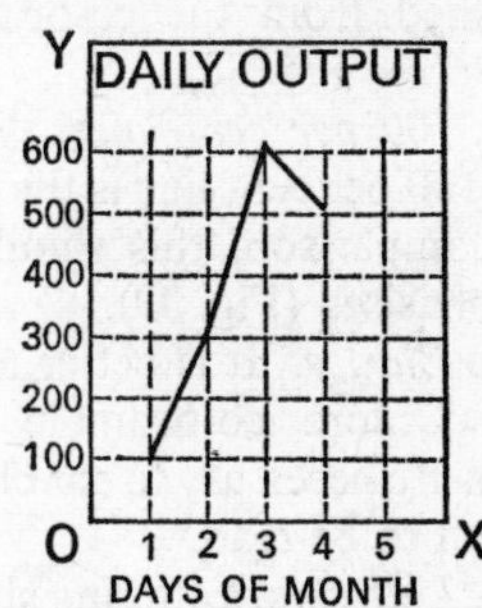

Note: In actual practice the letters YOX denoting the axes would not appear.

Fig. 36.—A historigram.

The point of intersection of the two axes is called the origin, denoted by O in the diagram. It will be observed that the co-ordinates form four quadrants. In business statistics only the quadrant YOX is usually of interest.

Having drawn the axes, and marked a suitable scale along both axes, the next stage is to plot the observations. This is best explained by means of examples. Suppose the value of $y = 4$ when the value of $x = 5$ and the value of $y = -3$ when the value of $x = -4$. These two observations are plotted as follows:

The first point is chosen at a distance of 5 (measuring along the x-axis) from the y-axis and at a distance of 4 (measuring along the y-axis) from the x-axis. Similarly, the second point is plotted at a position 4 units to the left of YOY_1 and 3 units below XOX_1 (see Fig. 35).

Rules for making graphs.

1. *Independent variable.* Decide which variable is the independent variable. In the case of a time series, time is the independent variable.

This is the x variable. In the case of a time series, the x-axis will, therefore, be scaled in years, months, etc.

2. *Zero line.* When the graph is concerned with absolute changes in quantities, the zero line must be shown, otherwise the graph will give a false impression (see Figs. 37 and 38).

3. *Co-ordinate lines.* The graph should be clearly distinguished from the co-ordinate lines (see Fig. 38).

4. *100 per cent lines.* When the 100 per cent line is the basis of comparison, this should be emphasised (Fig. 39).

5. *Clarity.* It is better not to show more co-ordinate lines than are necessary to enable the graph to be read.

6. *Lettering.* Lettering should be shown horizontally, not vertically.

7. *Scales.* The scales should be placed at the left and at the bottom of the graph; sometimes the vertical scale is repeated on the right (Fig. 40).

8. *Units.* Units should always be indicated at the top of the y-axis. The units relating to the horizontal scale are indicated as shown in Figs. 37, 39 and 40.

9. *Curves.* When more than one curve occurs on the same graph, these must be clearly distinguished.

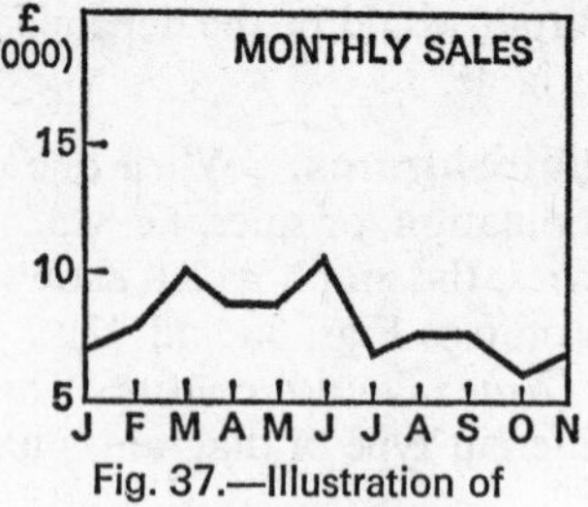

Fig. 37.—Illustration of Rules 2 and 8.

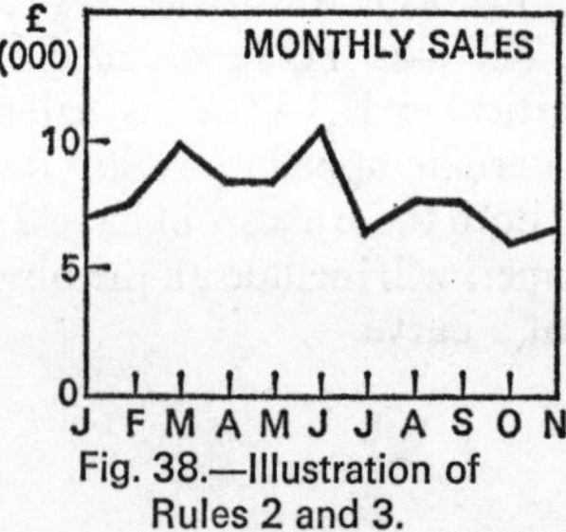

Fig. 38.—Illustration of Rules 2 and 3.

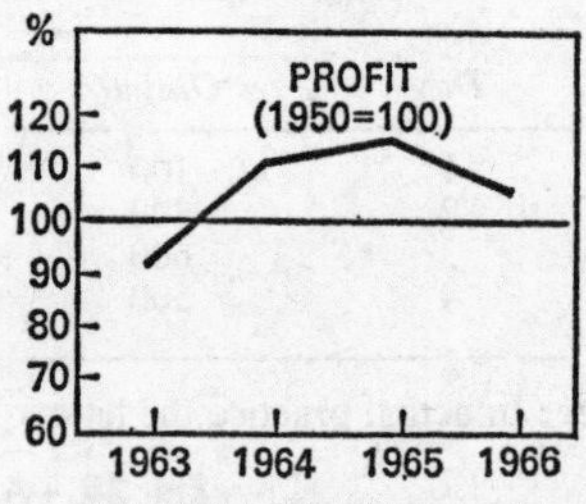

Fig. 39.—Illustration of Rules 4 and 8.

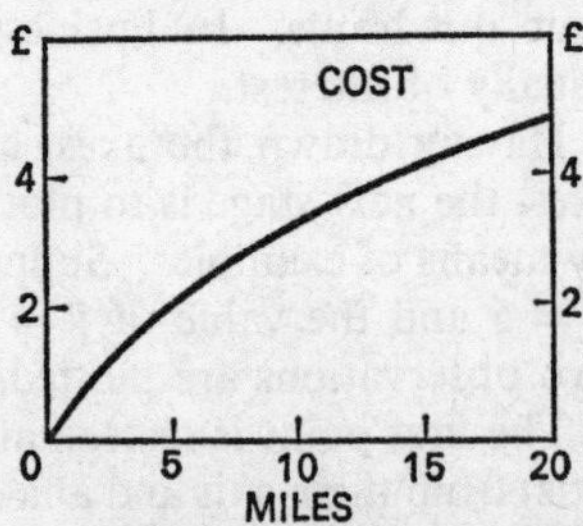

Fig. 40.—Illustration of Rules 7 and 8.

The most usual methods are:
(*a*) differentiated lines, *e.g.*

——— ——— – – – – · · · · · · — · —,

(*b*) different colours.

Both these methods necessitate a key.

(*c*) using captions as shown.

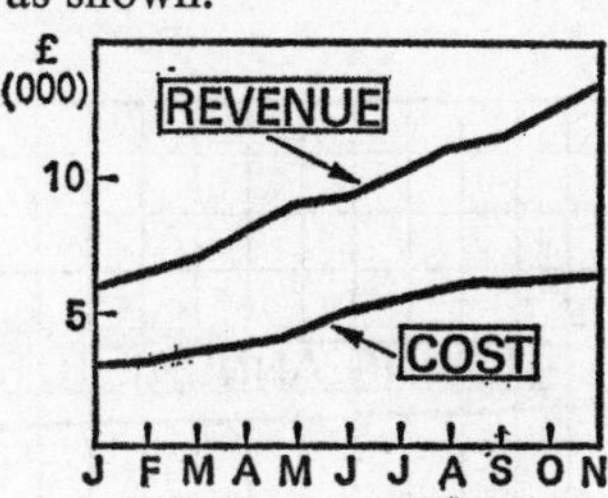

Fig. 41.—Illustration of Rule 9.

10. Too many curves should not be included in one graph.

11. *Title.* The graph must be given a title. This must be clear and concise. Where necessary, sub-titles and footnotes are added.

12. *Frame.* Often the graph is completely framed to give a finished appearance.

Examples of good graphs. Figs. 42 and 43 are excellent examples of the way in which graphs should be made. They are taken from *Economic Trends*, published by H.M.S.O. Note the differentiated lines to distinguish the years, the repetition of the vertical scale at the right in order to facilitate the reading of the graphs, how attention is directed to the fact that the space between the zero line and the first reading on the vertical scale is curtailed so that a false impression is not given. Note further sub-titles. In the case of Fig. 42, relating to production,

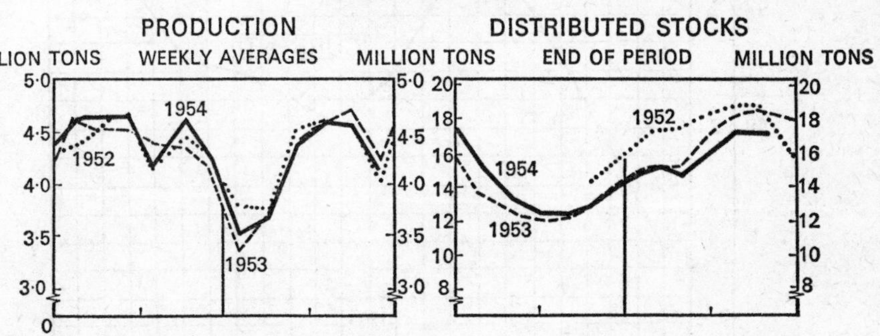

Fig. 42.—A graph with more than one curve.

Fig. 43.—A graph showing changes in stocks.

note that the points are plotted in the middle of the spaces relating to each month, whereas in the case of the graph relating to stocks the points are plotted at the end of the space relating to each month.

Whether production or sales are plotted in the middle of the spaces allotted to the periods to which they refer or whether they are plotted as in Fig. 36 is a matter of choice.

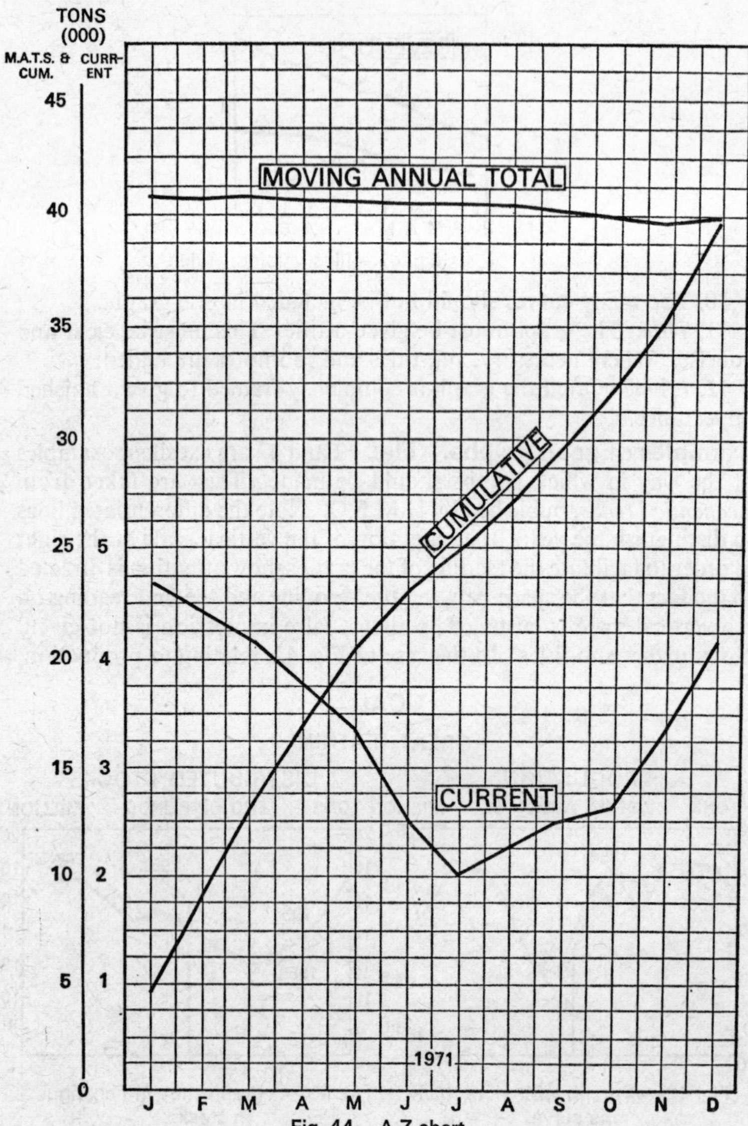

Fig. 44.—A Z chart.

The Z chart. This chart consists of three curves on a single chart. When drawn, the figure resembles the letter Z, hence its name. It is used to show movements in sales, production, etc.

The three curves are in respect of:

(*a*) the original data relating to each week, month or quarter as the case may be,
(*b*) the cumulative amount, and
(*c*) the moving annual total.

The scales used for the cumulative and annual total are usually smaller than the scale of the original data. This enables the movements of the original data to be shown more clearly. In the case of monthly figures, five times is a suitable amount. In the case of weekly figures, twenty times would be suitable. It is usual to differentiate the curve relating to the original data from the cumulative and the moving total curves by using different coloured lines. The cumulative curve always meets the moving total curve at the end of the year.

The curve relating to current data will show the fluctuations of a seasonal nature, the cumulative curve shows the position to date, and the moving total curve shows the trend.

Production

	Month	*Monthly (tons)*	*Cumulative total to date (tons)*	*Moving annual total (tons)*
1971	January	4,666	4,666	40,666
	February	4,416	9,082	40,582
	March	4,166	13,248	40,748
	April	3,966	17,214	40,714
	May	3,333	20,547	40,547
	June	2,500	23,047	40,547
	July	2,003	25,050	40,550
	August	2,247	27,297	40,297
	September	2,512	29,809	40,309
	October	2,688	32,497	39,997
	November	3,330	35,827	39,827
	December	4,169	39,996	39,996
1972	January	4,700	4,700	40,030
	February	4,515	9,215	40,129
	March	4,410	13,625	40,373

The year is from January to December. The cumulative totals will therefore re-commence each January. The cumulative total is the total to date; the cumulative total for April 1971 will therefore be $4{,}666 + 4{,}416 + 4{,}166 + 3{,}966 = 17{,}214$ as given in the table. The cumulative total for May 1971 will be $17{,}214 + 3{,}333 = 20{,}547$, again as given in the table. It is not possible to check the moving Annual

Total for 1971 as the previous year's monthly figures are not given. However, it is possible to check them for the three months of 1972. The moving total for December 1971 is 39,996, this being the total for the previous twelve months. To find the moving total for January 1972, the monthly figure for the previous January is deducted and the figure for January 1972 is added, thus 39,996 − 4,666 + 4,700 = 40,030. For February 1972 the total for the twelve months ended January 1972 is taken, the production of the previous February subtracted and the production for February 1972 added, thus 40,030 − 4,416 + 4,515 = 40,129.

The Z chart for 1971 is shown in Fig. 44.

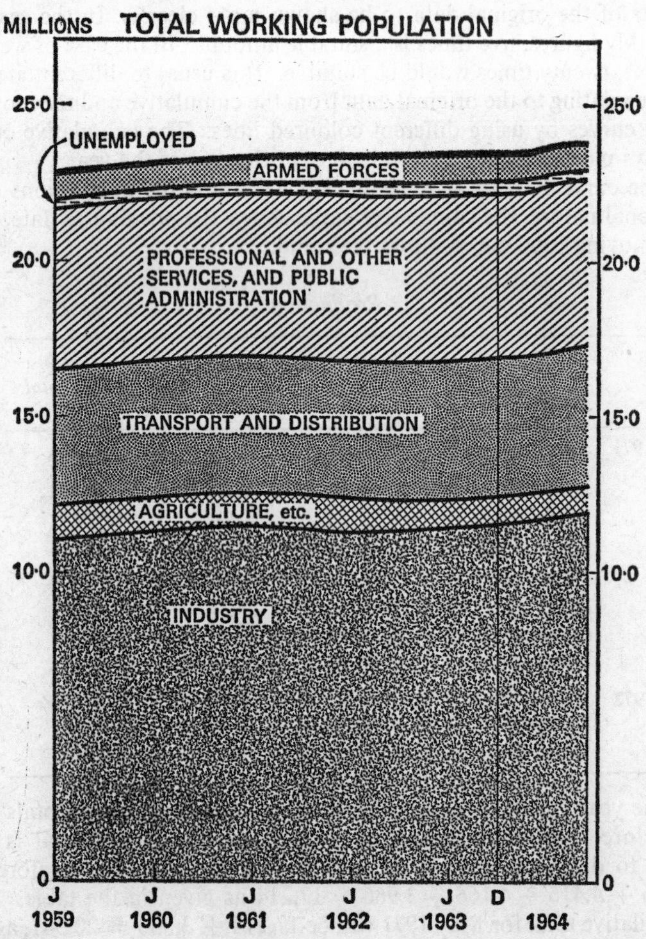

Fig. 45.—A band-curve chart showing composition of total working population by employment.

Band-curve charts. In these charts, the constituent parts of the whole are plotted one above the other, so that the charts appear, when shaded, to consist of a series of bands.

The whole space between the top curve and the base is filled in solidly (Fig. 45). Band-curve charts can also be made where the data are put into percentage form; the whole chart will depict 100 per cent and the bands the percentage each component bears of the whole.

The Lorenz Curve. This curve is a graphic method of showing to what extent various magnitudes vary from uniformity.

Examples of the use of such a curve are:

(*a*) to show the degree of concentration of a particular industry in the hands of a few firms;

(*b*) the extent to which the incomes of various groups of people vary;

(*c*) the extent to which the profits of various firms vary.

The basis of the curve is the plotting of the cumulative percentage of the number of items, *e.g.* firms or people against the cumulative percentage of, for example, the sales, incomes or profits distributed among those items.

If all the firms in a particular industry were the same size (measuring size by their sales), then 10 per cent of the firms would have 10 per cent of the sales, 30 per cent of the firms would have 30 per cent of the sales, and so on. Plotting cumulative percentage sales against cumulative percentage number of firms would result in a straight line, and since the chart is square, the line would be at an angle of 45°. This is the line of equal distribution and is always drawn. The more the actual curve diverges from this line the greater the divergence from a uniform distribution.

Stages in the drawing of a Lorenz curve.

(*a*) Draw up a frequency table as in the first two columns of Fig. 46.

(*b*) Compute the amount of sales, profits, incomes as the case may be for each group. In Fig. 46, this is savings and is given in column 3.

(*c*) Compute the cumulative amounts, in the example shown, savings and the cumulative frequencies (columns 4 and 6 in Fig. 46).

(*d*) Compute the cumulative percentage amounts and the cumulative percentage frequencies (columns 5 and 7 in Fig. 46).

(*e*) Plot cumulative percentage frequencies against cumulative percentage amounts.

(*f*) Draw the 45° diagonal.

(*Note:* The scales are the same for both axes.)

Lorenz Curve

Annual savings	Fre-quency	Amount saved	Cumulative savings		Cumulative frequency	
		£	£	%		%
Not exceeding 50p	9,133	2,511	2,511	29	9,133	80
50p to £1·00	1,503	998	3,509	41	10,636	93
£1·00 to £5·00	546	1,002	4,511	53	11,182	98
£5·00 to £15·00	95	971	5,482	64	11,277	99
£15·00 to £35·00	51	1,024	6,506	76	11,328	99
£35·00 to £100·00	22	985	7,491	89	11,350	99
Over £100·00	5	1,013	8,504	100	11,355	100
	11,355	8,504				

Fig. 46.—A Table showing data and computations to make a Lorenz chart.

Interpretation of a Lorenz Curve. Fig. 47 shows that the greater amount of savings is done by comparatively few people; that a large number of people only save a comparatively small amount of the total. In other words, there is a great divergence between the amounts saved by the various savers.

If more than one curve is plotted on the same chart, it is possible to compare, say, whether teachers' salaries vary more than those of bank clerks, or whether there is a greater concentration of trade in fish shops

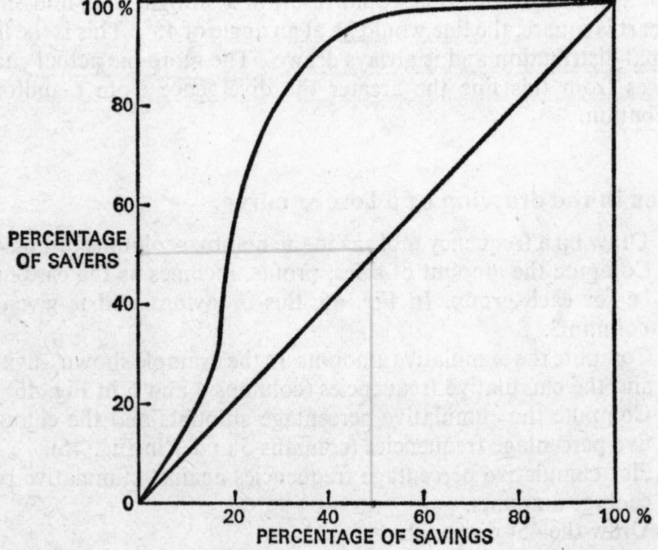

Fig. 47.—A Lorenz curve.

or butchers' shops. The farther away from the diagonal, the greater the amount of concentration. It is also possible to make comparisons between areas, or at different times.

It is, of course, equally possible to make the comparisons by comparing different charts. However, it is easier if the curves are on the same chart.

Questions

1. From the sales of Puro-Products Ltd. given below, draw a Z chart.

	1970 (£000s)	*1971 (£000s)*		*1970 (£000s)*	*1971 (£000s)*
January	143	156	July	208	306
February	172	182	August	135	214
March	181	201	September	118	239
April	209	234	October	241	337
May	412	512	November	326	385
June	413	511	December	389	411

2. From the data given below, draw a band-curve chart.

Road Casualties

	Total	*Pedestrians*	*Pedal cyclists*	*Motor-cyclists and passengers*	*Other drivers and passengers*
1967	216,493	59,861	48,077	42,680	65,875
1968	208,012	54,503	47,244	42,644	63,621
1969	226,770	58,553	49,618	49,438	69,161
1970	238,281	61,381	49,295	52,531	75,074
1971	267,922	63,897	52,447	63,602	87,976

3. State briefly, giving reasons, the kind of diagram you consider most appropriate for use with each of the following classes of statistical data:

(*a*) Number of children per family in a large town.
(*b*) Monthly rainfall for a period of three years.
(*c*) Total ship canal dues for one year according to country of registration of ships using the canal.
(*d*) Noon temperature for each day for six months at one weather station.
(*e*) Monthly output of steel for one year according to the principal grades of quality.

Chartered Institute of Transport.

4. What is meant by a historigram? From the data below, draw a historigram.

Gross National Product

£ million

1962 . .	8,787	1967 . .	12,793
1963 . .	9,387	1968 . .	13,928
1964 . .	10,376	1969 . .	14,858
1965 . .	11,057	1970 . .	15,909
1966 . .	11,636	1971 . .	16,784

5. The Managing Director of a firm engaged in the manufacture of men's shirts asks you to produce information in the form of tables or graphs, which he wishes to see as a matter of routine. Draft four specimen tables or graphs stating how often they would be compiled and what information he would derive from each of them. *Institute of Statisticians.*

6. What does a graph show? On which axis is the independent variable plotted?

7. Draw up a list of rules for the construction of graphs.

8. The following figures come from a Report on the Census of Production.

Textile machinery and accessories

Establishments	*Net output*
Nos.	£000
48	1,406
42	2,263
38	3,699
21	2,836
26	3,152
16	5,032
23	20,385
214	38,773

Analyse this table by means of a Lorenz curve and explain what this curve shows. *Institute of Cost and Management Accountants.*

CHAPTER 13

RATIO-SCALE GRAPHS

The graphs in the preceding chapter were concerned with movements in absolute quantities, *e.g.* an increase or a decrease of so many tons or so many £s. In the present chapter the movements that are to be shown are not the absolute movements, but the rates of change of the movements, *i.e.* relative changes.

Relative and absolute changes. The difference is best explained by means of an example.

Consider a retail shop whose sales are in four successive weeks as follows:

£125 £150 £180 £216.

Successive increases are:

£25 £30 £36.

These are absolute quantities.

Percentage increases are as follows:

$$\frac{25 \times 100}{125}\% \qquad \frac{30 \times 100}{150}\% \qquad \frac{36 \times 100}{180}\%.$$

These are relative changes. It will be noted that the increase in sales over the preceding week is in each case 20 per cent. If the original data were plotted on semi-logarithmic paper, the result would be a straight line, showing that the rate of increase was constant. On ordinary natural-scale graph paper absolute differences would be shown and it would not be possible to say anything about the rate of change.

The uses of semi-logarithmic scale graphs. If the interest is in the rate of change of data, and not in the actual absolute changes, the data must be plotted on semi-logarithmic paper; such graphs are known as ratio-scale graphs or semi-logarithmic graphs.

Here are some of the uses of such a graph.

(*a*) Comparison of growth of a firm's sales with growth of total sales of the industry to see if the firm is obtaining its share of sales.

(*b*) Comparison of the efficiency of salesmen.
(*c*) Comparison of outputs of various departments.
(*d*) Estimation of future sales.
(*e*) Comparison of the fluctuations of prices of different commodities.

How to use semi-logarithmic paper. On examining semi-logarithmic paper, a number of differences from ordinary natural-scale paper will be noticed: there is no zero line on the vertical scale, the lines from 1 to 10 become closer and closer, the pattern is repeated, but the values are ten times greater. If the pattern were repeated a third time, *i.e.* if the paper had three cycles, the third cycle would be from 100 to 1,000, *i.e.* ten times greater than the second cycle. Similarly the fourth cycle would have values ten times greater than the third, and so on. Alternatively, the first cycle could have started with the value of 0·1 and finished at 1, in which case the second cycle would have values from 1 to 10, the third 10 to 100, and so on. Note that the units are sub-divided into ten parts when the lines are fairly far apart and into 5, when they are closer together. Note further that the sub-divisions also become closer and closer. This scale is logarithmic. The horizontal scale is the ordinary arithmetical scale. Hence the paper is known as semi-logarithmic.

Sales

	Firm (£000)	Industry (£m)
1965	1·50	3·2
1966	2·00	4·0
1967	2·18	5·4
1968	2·69	6·8
1969	3·50	8·0
1970	6·00	11·0
1971	14·00	19·0

Fig. 48.—A table showing a firm's share of trade.

Actual values are plotted on this paper. If, however, semi-logarithmic paper is not available, the logarithms of the values can be plotted on arithmetically scaled paper, but the actual values are inserted on the scale. This is, in effect, making a logarithmic scale.

The reason for the scale on semi-logarithmic paper starting at 1, and not zero, is that the logarithm of 1 is 0; hence the value of 1 is placed at zero distance from the origin, *i.e.* at the origin. There is no logarithm for zero, nor for negative numbers, hence such values cannot be plotted (see Chapter 7).

The students should carefully note exactly how the values given in Fig. 48 are plotted on the ratio-scale graph on Fig. 49.

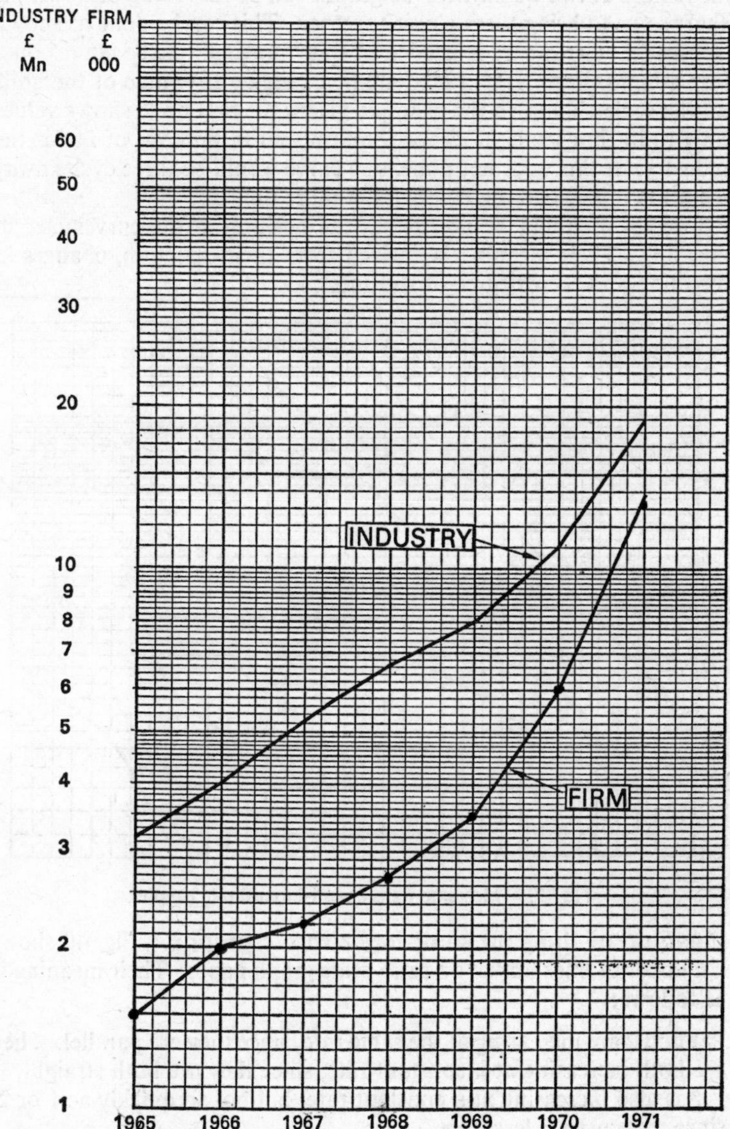

Fig. 49.—Comparison of relative movements on ratio-scale graphs.

In order to make comparisons easier when comparing the rates of movement of two quantities, the graphs can be moved up or down the chart so as to bring them closer together. This can be done as shown in Fig. 49. Great care is needed in this "moving" of the curves on a logarithmic chart. It can be done by changing the value of the units on the scale. Thus, in the example given, the first cycle shows values 1 to 10 in thousands of £s for the firm, but in millions of £s for the industry. If this were not done, one curve would be three cycles away from the other, making comparison more difficult.

Finally, it should be noted that the position of the curves has no meaning. Their meaning is derived from their direction, changes in direction and their degree of steepness.

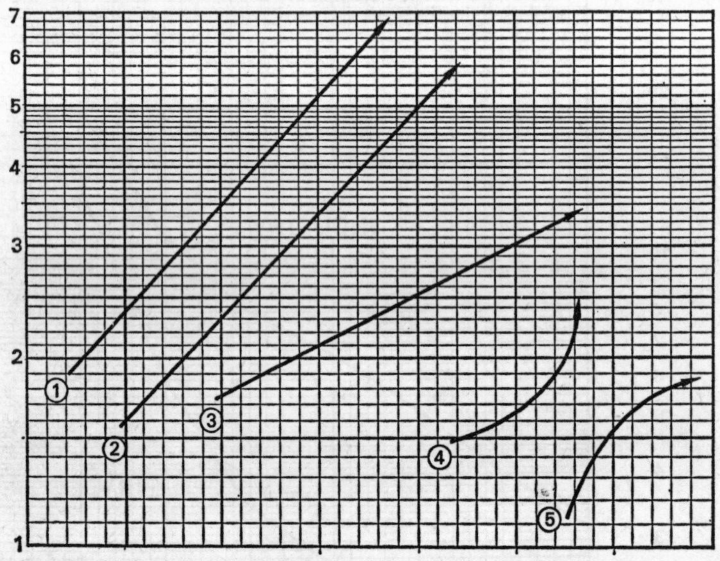

Fig. 50.—Movements on semi-logarithmic graphs.

Meaning of lines on semi-logarithmic graphs. Fig. 40 shows a number of lines drawn on ratio-scale graph paper. Their meaning is as follows:

1 and 2 are increasing at the same rate, since they are parallel. They are both increasing at a constant rate, since they are both straight.

3 is also increasing at a constant rate, but not so quickly as 1 or 2, since this curve is less steep.

4 is increasing at an increasing rate, since it is becoming steeper and steeper.

5 is increasing at a decreasing rate, since the curve is becoming less and less steep.

A comparison of ratio-scale and arithmetical-scale graphs.

A comparison can be made of these two types of graph as follows:

Ratio-Scale	*Arithmetical-Scale*
(*a*) Used to show relative movements.	(*a*) Used to show movements of absolute quantities.
(*b*) Equal vertical distances represent equal proportional changes. (If 1 inch represents an increase of 40 per cent in one part of the graph, then 1 inch represents 40 per cent in any other part. However, ½ inch does not represent 20 per cent.)	(*b*) Equal vertical distances represent equal amounts. (If 1 inch represents £4, then 1 inch represents £4 in any other part of the graph. ½ inch will represent £2.)
(*c*) Zero and negative quantities cannot be shown.	(*c*) Zero and negative quantities can be shown.
(*d*) No base line necessary. Graph can be moved up and down without changing meaning.	(*d*) Base line necessary. Graph cannot be moved up or down without destroying its meaning.
(*e*) Meaning derived from direction of lines.	(*e*) Meaning derived from position of lines.
(*f*) Cannot be used to show component parts.	(*f*) Can be used to show components (band-curve charts).
(*g*) Two or more series of entirely different units, *e.g.* tons, gallons, £s can be compared on the same graph.	(*g*) It is not possible to show two or more series on the same graph for purposes of comparison if they are of entirely different units.
(*h*) A very great range of values can be shown.	(*h*) Cannot show a very great range of values if the graph is to be clear.

Questions

1. Discuss the advantages and disadvantages of logarithmic charts. Describe and illustrate two practical applications of this type of diagram.

Institute of Company Accountants.

2. The following table shows the number of television licences current at the end of the stated periods.

		000's
1948		93
1949		239
1950		578
1951	March	764
	June	897
	September	949
	December	1,162
1952	March	1,456
	June	1,539
	September	1,655
	December	1,839
1953	March	2,142
	June	2,415
	September	2,615

Plot the series on a suitable graph and estimate the figure as at end December 1953. Would this series be useful for the purpose of interpreting trends in passenger traffic?

Chartered Institute of Transport.

3. Discuss the limitations and uses of graphs in statistical analysis. State the respective merits of (*a*) natural-scale graphs; (*b*) semi-logarithmic graphs; (*c*) bar charts.

4. Explain what is meant by a semi-logarithmic diagram and discuss its advantages over the natural-scale diagram.

Graph the two following series on the same diagram so as to bring out their relative movements.

New Factory Buildings in Patonia

Number Completed

		Total	*No. in development areas*
1969	1 . .	264	92
	2 . .	267	122
	3 . .	215	48
	4 . .	276	90
1970	1 . .	256	58
	2 . .	284	65
	3 . .	218	48
	4 . .	264	48
1971	1 . .	241	59
	2 . .	248	65
	3 . .	271	46
	4 . .	279	51

London Chamber of Commerce and Industry.

5. What are the respective merits of (*a*) arithmetic charts, (*b*) semi-logarithmic charts and (*c*) bar charts, in presenting statistics showing (i) sales and (ii) stocks of a number of different products? Indicate, with reasons, which method you regard as best for each set of statistics. *Institute of Statisticians.*

6. State the chief points to be considered in the construction of (*a*) the arithmetic-scale graph and (*b*) the ratio-scale graph. Graph the data below using a ratio-scale graph.

Road Accident Deaths

	1966	*1967*	*1968*	*1969*	*1970*
Number of deaths per million of population	106	102	93	98	102
Consumption of motor fuel in 10 million gallons . .	35	31	30	29	27

7. Below are given estimates of world population at various dates. From this information make an estimate of the population for the year 2020, using graphical means.

Year	*1670*	*1770*	*1870*	*1920*	*1970*
Population (millions)	470	694	1,091	1,550	2,406

8. Using only the following information:

$$\log 2 = 0{\cdot}301, \log 3 = 0{\cdot}477$$
$$\log 7 = 0{\cdot}845$$

illustrate clearly how you would rule some semi-logarithmic paper, using ordinary graph paper. Show fully how you would insert scales on both axes.

Institute of Cost and Management Accountants.

CHAPTER 14

FREQUENCY DISTRIBUTIONS

Variables. When a quantity varies from one individual to another it is known as a variable or variate. Examples are the prices of a given commodity from shop to shop, the earnings of employees within a firm, the sales from month to month, production from year to year; the list is endless.

Variables are of two kinds: those which can take any value, *e.g.* the weights of people—these are continuous variables—and those which can take only particular values, *e.g.* the number of children in a family (it is not possible to have three-and-a-half children)—these variables are known as discontinuous or discrete variables.

Frequency distributions. These give the number of items (the frequency) of each size (the variable) of the subject-matter of our investigation. Usually, sizes are put into groups, even when the variables are discrete; for example:

0–2, 3–5, 6–8 children per family.

Frequency tables. These give frequency distributions in the form of tables.

Example 1.

Weekly Wage	*No. of Employees*
£	
Under 5	17
5 and under 8	23
8 " 10	51
10 " 15	9
	100

The number of employees in each group is the frequency. It means that 17 employees earn under £5, 23 earn between £5 and £8, etc. The frequencies of each class when totalled amount to the total number of items—in this case, 100. In Example 1, the groups, known as classes, have as class frequencies 17, 23, 51 and 9. The intervals—the differences between the highest and lowest values in each class—are known as class intervals. They are £5, £3, £2 and £5 respectively. In Example 1, the class intervals are unequal. The variable is the weekly wage.

Compiling frequency tables. The tables are compiled from a mass of original data, consisting of items of specific measurements. Two problems require solution. How many groups? What class intervals or, more specifically, what shall be the limits of the class intervals? The grouping must be so done that the arithmetic average as computed from the frequency table shall, for all practical purposes, be the same as the arithmetical average which would be calculated from the individual items. Further, if at all possible, class intervals should be made equal.

If too great a number of classes are chosen, certain classes may have too few frequencies and the distribution may be shown as more irregular than it really is. On the other hand, if too few classes are chosen, the nature of the distribution will be more approximate than is necessary. This point will become clearer when frequency distributions are shown graphically, later in this chapter. The greater the number of items to be grouped, the greater the number of classes. A maximum of 20 classes when there is a very great amount of data, and a minimum of, say, 5 when the number of items is small might be considered as a practical rule. It is necessary to be able to treat all the values in any one class as if they were all equal to the mid-value of the interval. In Example 1, the 51 employees earning between £8 and £10 should earn on the average £9 per week. Subject to this over-riding consideration, the class intervals should be as large as possible.

The selection of the class limits is made to suit the convenience of the compiler. He will wish to avoid any difficulty in assigning any particular item to a particular class; in order to do this he will, for example, avoid making class limits coincide with recorded values. Values around which a large number of items occur will fall towards the mid-values of intervals.

Mathematical limits of class intervals. Below are four ways in which the limits of class intervals are shown.

The first three cases are different ways of expressing the same limits, the assumption being that the measurements, for example, the under 10 acres column, are taken to a sufficiently precise degree to warrant including items which are for all practical purposes equivalent to 10 acres, *e.g.* 9·9897 acres. The limits shown in case C are known as mathematical limits, the highest value of one class being the lowest value of the next class.

Case D varies from the other three. In the absence of information to the contrary, the assumption made is that the measurements are

Example 2.

A (acres)	B (acres)	C (acres)	D (acres)
5–	5 and under 10	5–10	5–9
10–	10 " 15	10–15	10–14
15–	15 " 20	15–20	15–19
20–	20 " 25	20–25	20–24
25–	25 " 30	25–30	25–29

recorded to the nearest acre. The mathematical limits are therefore as follows: $4\frac{1}{2} - 9\frac{1}{2}$; $9\frac{1}{2} - 14\frac{1}{2}$, etc.

In the case of ages, this assumption is not valid. Ages are usually recorded as age last birthday. This is so for ages returned by the Registrars-General. Hence:

Years		*Years*
0–4		0–5
5–9	are treated as	5–10
10–14		10–15.

In all cases, mathematical limits must be used for calculations and in making graphs of the distribution.

Histograms. A frequency distribution can be shown in the form of a diagram. The diagram in Fig. 51 is known as a histogram (also called a block frequency diagram). It shows the pattern of the distribution, whether, for example, it is symmetrical.

Frequency Distribution

Ounces	*Number*
3–6	2
6–9	5
9–12	7
12–15	2
	16

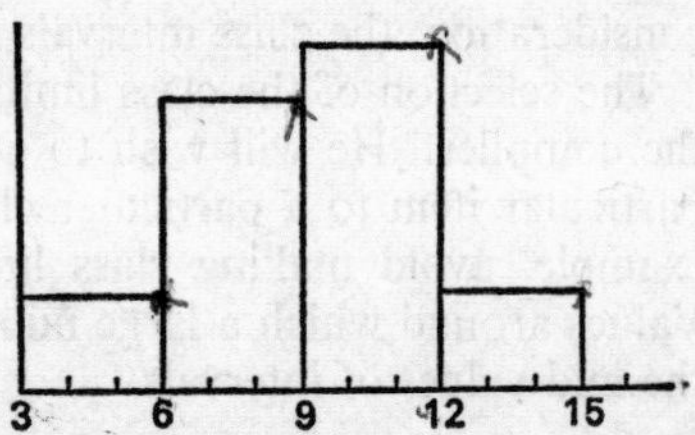

Fig. 51.—A histogram with equal class intervals.

The class intervals are measured along the horizontal axis. Rectangles are erected whose AREAS ARE PROPORTIONAL TO THE FREQUENCIES. Where the class intervals are equal, the heights of the rectangles will be proportional to the frequencies, but the basic principle to grasp is that the areas are in any case proportional to the frequencies.

If the intervals are unequal, the heights of the rectangles will be the frequency (the area) divided by the class interval (the width). Frequency divided by class interval is known as *frequency density*.

Frequency Distribution

Ounces	*Number*	*Frequency density*
2–6	8	$8 \div 4 = 2$
6–9	9	$9 \div 3 = 3$
9–14	10	$10 \div 5 = 2$

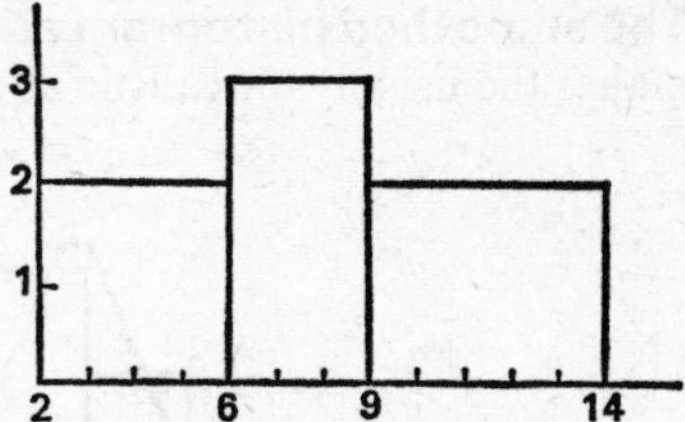

Fig. 52.—A histogram with unequal class intervals.

Frequency polygons. Frequency distributions can also be shown in diagrammatic form by means of frequency polygons. A frequency polygon is obtained by joining the mid-points of the tops of the rectangles of a histogram. An example is given in Fig. 53. Note the treatment for the first and last rectangles.

The polygon thus drawn will have the same area as the corresponding histogram, but only if the class intervals are the same.

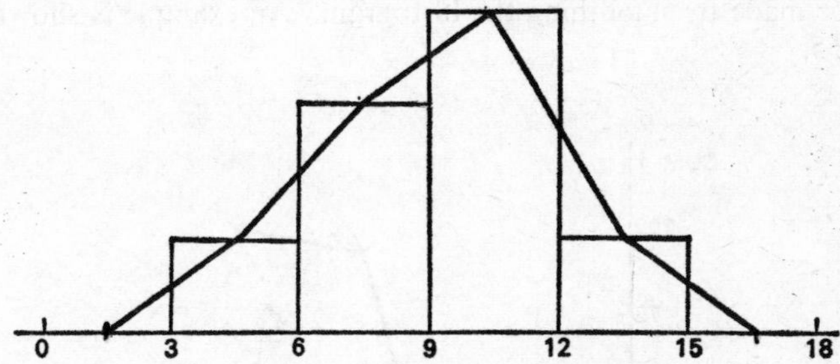

Fig. 53.—A frequency polygon.

Frequency curves. As the class intervals are made smaller and smaller and the number of observations increase, the frequency polygon and the histogram approach a smooth curve. If this process were carried on to the limit, the result would be a frequency curve.

Notes: The shaded area is proportional to the number of observations whose size vary between x_0 and x_1. The number of observations smaller in size than x_0 will be proportional to the area to the left of x_0.

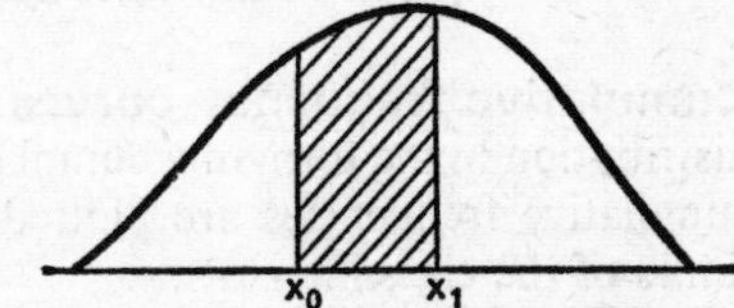

Fig. 54.—A frequency curve.

The area inside the curve will be proportional to the total number of observations, *i.e.* the total of the frequencies.

The smoothed histogram as a frequency curve. The histogram makes the usually unrealistic assumption that over the class intervals

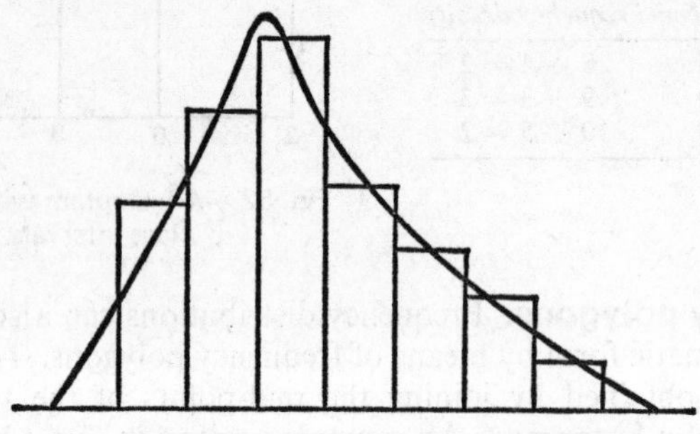

Fig. 55.—A "smoothed" histogram.

different magnitudes have the same frequencies. For this reason a frequency curve is more accurate. An attempt to obtain such a curve can be made by smoothing the histogram. An example is shown in Fig. 55.

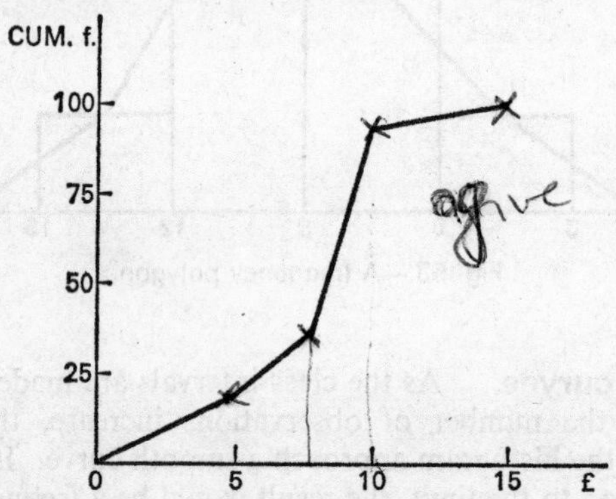

Fig. 56.—A cumulative frequency curve (or Ogive).

Cumulative frequency curves. In order to present a frequency distribution in the form of a cumulative frequency curve (see Fig. 46), cumulative frequencies are plotted against the upper mathematical limits of the class intervals.

This means that 17 items do not exceed £5, 40 items do not exceed £8, and so on.

Unequal intervals cause no difficulty, cumulative frequencies are plotted against the upper mathematical limits in all cases. It will be

Variable weekly wage	Frequency	Cumulative frequency
£		
Under 5	17	17
5 and under 8	23	40 (17 + 23)
8 ,, 10	51	91 (40 + 51)
10 ,, 15	9	100 (91 + 9)

noted that the points have been joined by straight lines. This is usual, although inaccurate; a smooth curve should pass through the points.

Cumulative percentage frequency curves. In making such a curve, the cumulative percentage frequency is plotted against the values of the upper limits of the class intervals. The cumulative per-

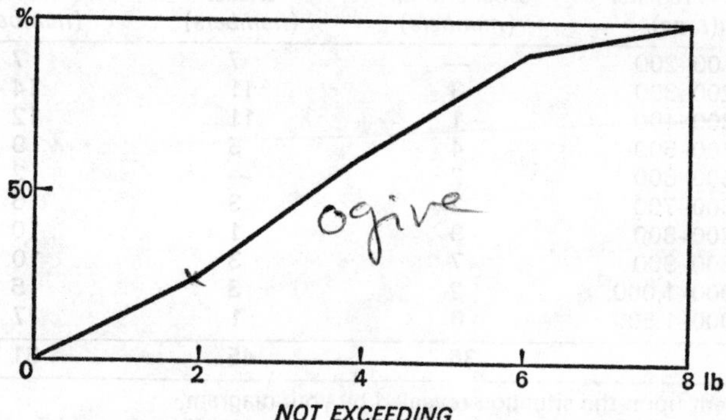

Fig. 57.—A cumulative percentage frequency curve.

Weight (lb)	Frequency	Cumulative frequency	Percentage cumulative frequency
0–2	5	5	$\frac{5}{20} \times 100 = 25\%$
2–4	7	12	$\frac{12}{20} \times 100 = 60\%$
4–6	6	18	$\frac{18}{20} \times 100 = 90\%$
6–8	2	20	$= 100\%$

centage frequencies are the cumulative frequencies expressed as percentages of the total number of observations.

When it is required to make a comparison on the same diagram of two or more cumulative frequency curves, cumulative percentage curves must be used (see Fig. 57).

Questions

1. (*a*) Define the statistical term "histogram."

(*b*) Explain the method of constructing histograms when the class intervals are unequal.

(*c*) In a savings group there are 400 members and the number of savings certificates held by them is shown in the following table:

No. of certificates held	50 and under	51–100	101–150	151–200	201–300
No. of members	10	15	30	40	120
No. of certificates held	301–400	401–500	Total		
No. of members	100	85	400		

Construct a histogram showing the distribution of savings certificates.

2. Show the following three distributions as frequency polygons on one diagram:

Coasting and Home Trade Tankers

Gross register (tons)	*Coal and oil (numbers)*	*Diesel (numbers)*	*Total (numbers)*
100–200	—	7	7
200–300	3	11	14
300–400	1	11	12
400–500	4	5	9
500–600	2	—	2
600–700	2	3	5
700–800	9	1	10
800–900	7	3	10
900–1,000	2	3	5
1,000–1,500	6	1	7
	36	45	81

Comment upon the situation revealed by your diagram.

Chartered Institute of Transport.

3. Describe each of the diagrams listed below and give an illustration in each case.

Bar diagram, compound bar diagram, multiple bar diagram, percentage bar diagram, block frequency diagram, ogive, Lorenz Curve, logarithmic chart, line chart. *Institute of Company Accountants.*

4. *England and Wales—Resident Population*
31st December 1951

Age groups	(*Thousands*) *Males*	*Females*
0–4	1,853	1,768
5–14	3,133	3,004
15–24	2,709	2,864
25–34	3,132	3,200
35–44	3,284	3,377
45–54	2,928	3,135
55–64	2,058	2,559
65 and over	1,988	2,885

Source: *Registrars-General.*

Draw diagrams of both age distributions, given above, in frequency blocks. Use your diagrams to contrast and compare the variations in the two distributions.

5. The following table shows the number of operators of public service vehicles in the Metropolitan Traffic Area (excluding L.T. executive), grouped according to the number of vehicles owned by each operator:

Vehicles owned as at 31.3.1948	*Number of operators*
1	85
2	51
3	32
4	29
5–9	81
10–19	55
20–49	8
50–99	8
100–199	1

Show this distribution in the form of a histogram.

Chartered Institute of Transport.

***Note:* Give vehicles owned mathematical limits, viz: $\frac{1}{2}$–$1\frac{1}{2}$; . . . $4\frac{1}{2}$–$9\frac{1}{2}$; etc.**

6. The following table relates to the entries to the Intermediate Administrative Examinations for two years. Make a comparison between the two distributions by means of cumulative frequency curves.

Ages	*1953*	*1954*
19	9	9
20–24	188	198
25–29	161	208
30–34	123	183
35–39	67	82
40–44	14	40
45–49	3	15
	565	735

7. Indicate the more important uses of any two of the following charts and graphs, and provide diagrammatic examples to support your case:

(*a*) Cumulative frequency curve
(*b*) Z (or Zee) chart
(*c*) Ratio scale chart
(*d*) Ideograph (pictogram)

I.C.S.A. Part 1, Dec. 1975.

CHAPTER 15

THE ARITHMETIC MEAN AND THE HARMONIC MEAN

The meaning of an average. In any given distribution of sizes there is some size, not necessarily coincident with any value in the distribution, around which the other values are distributed. The value around which the other values lie is a measure of a central tendency, a measure of location. This central value is called an average.

Kinds of average. There are four kinds of average (sometimes called a mean) which are commonly used:

(*a*) The Arithmetic Average or Arithmetic Mean.
(*b*) The Geometric Average or Geometric Mean.
(*c*) The Mode.
(*d*) The Median.

The arithmetic average or mean. This is the sum of a group or series of items divided by their number.

The simple arithmetic average is calculated thus: the numbers of orders taken by 10 salesmen on July 31st were: 9, 10, 5, 6, 9, 12, 7, 8, 8, 6. The total orders were 80.

The simple arithmetic average is $\frac{80}{10} = 8$ orders per man.

The important point is that all the units must be homogeneous for the purpose under consideration.

A short method of averaging large numbers. This is as follows: Assume an average by inspection, find the deviation of number from this assumed average, take the average of the deviations, and add to or subtract from the assumed average. The result is the true average. Deviations are calculated by deducting assumed average from values of items (see Fig. 58).

Weighted arithmetic average. In the simple average, each item in the averaged group is regarded as of equal importance. Items vary in importance, however, and to obtain a representative average it is necessary to multiply each item by a suitable "weight" corresponding

Calculating Averages

Items	Assumed average	Deviations
853	850	+3
845	850	−5
857	850	+7
862	850	+12
841	850	−9
859	850	+9
861	850	+11
		Total +28

Number of items = 7. Average of deviations $28 \div 7 = 4$.
True Simple Average $850 + 4 = 854$.

Fig. 58.—A short method of averaging large numbers.

to its importance, and divide the total of the products by the sum of the weights. This is then called a weighted arithmetic average. The procedure is better shown by Fig. 59.

Weighted average using approximate weights. When actual numbers of items can be used as weights, this is best, but for some purposes when actual quantities are not known, then a close estimate may be made, and the estimated numbers can be used as weights. Estimated or approximated weights, if reasonably selected, make very little error in the resultant average (see Fig. 60).

Daily Wages Paid by P.Q.R., Ltd.

Daily wages *(a)*	No. of workers at each rate of wage *(b)*	Product *(a)* x *(b)*
£		£
3·00	90	270
3·50	120	420
4·00	590	2,360
5·00	420	2,100
6·00	130	780
21·50	1,350	5,930

$$\text{Weighted average} = \frac{5{,}930}{1{,}350} = £4{\cdot}39$$

The simple average would be $\frac{£21{\cdot}50 \text{ (total of column } (a))}{5} = £4{\cdot}30$ per week, but the weighted average gives an average daily wage of £4·39 through taking account of the bigger proportion of workers at £4 and £5.

Fig. 59.—A weighted arithmetic average.

Daily Wages Paid in P.Q.R., Ltd.

Daily wages	Estimated no. of each wage	Weight	Product
£			£
3·00	150	1·5	4·5
3·50	150	1·5	5·25
4·00	600	6·0	24·00
5·00	400	4·0	20·00
6·00	150	1·5	9·00
		14·5	62·75

$$\text{Weighted average} = \frac{62{\cdot}75}{14{\cdot}15} = £4{\cdot}33$$

Fig. 60.—A weighted average using estimated weights.

It will be noticed that although the estimated figures differ considerably from the actual in Fig. 59, the average is £4·33, as compared with £4·39 using actual numbers.

The following example shows the importance of using weighted values.

Coffee is sold at 40p, 50p, 80p, 120p and 160p per lb. The simple arithmetic average is:

$$\frac{40\text{p} + 50\text{p} + 80\text{p} + 120\text{p} + 160\text{p}}{5} = \frac{450\text{p}}{5} = 90\text{p per lb.}$$

This figure is only of value if an equal weight of each grade were sold. In fact sales were 1,362 lb, 1,200 lb, 961 lb, 450 lb and 222 lb respectively. These quantities must be used as weights:

$$\frac{(1{,}362 \times 40\text{p}) + (1{,}200 \times 50\text{p}) + (961 \times 80\text{p}) + (450 \times 120\text{p}) + (222 \times 160\text{p})}{1{,}362 + 1{,}200 + 961 + 450 + 222}$$

$$= \frac{280{,}880}{4{,}195} = 67\text{p per lb.}$$

The work can be reduced by using approximate relative weights:

$$\frac{(14 \times 40\text{p}) + (12 \times 50\text{p}) + (10 \times 80\text{p}) + (5 \times 120\text{p}) + (2 \times 160\text{p})}{14 + 12 + 10 + 5 + 2}$$

$$= \frac{2{,}880}{43} = 67\text{p per lb.}$$

Advantages and disadvantages of the arithmetic average.

1. *Advantages.*

(*a*) It is easy to understand and calculate, and is commonly used.

(*b*) It makes use of all the data in the group, and can be determined, therefore, with mathematical exactness.

(*c*) It can be determined when nothing more than the total value or quantity of the items, and the number of them, are known.

2. *Disadvantages.*

(*a*) It may give undue weight to, and be unduly influenced by, extreme abnormal items, *e.g.* in a business there may be a few earning very high wages, making the average higher than it would be without these included, and although the bulk of the workers are earning the same wage as those in another business, yet the average would give the impression, wrongly, that better wages are being paid in the first business.

(*b*) The average may be a value which does not correspond with a single item, *e.g.* in Fig. 59, for instance, no worker earns exactly £4·39 a day. The average number of employees in each department of a business may be found to be 4·86 persons, but obviously the actual number in each must be an exact whole number.

The arithmetic average is the most useful average for general purposes, and should be used unless particular reasons call for another method.

Uses of the arithmetic average. In business this is useful in several ways:

- Revenue and expenses.
- Sales, daily, monthly, etc.
- Output (weekly, etc.).
- Output (per man, per machine, etc.).
- Number of employees by sexes, per dept., etc.
- Plant in service.
- Departmental, etc. allocation of fixed expenses, *e.g.* rent by floor space.
- Operation costs.
- Sales per article.
- Rates of pay.
- Piece-work earnings.
- Bonus earnings.
- Per capita consumption of articles.
- Labour and machine hour rates for recovering on-cost.

Arithmetic mean of grouped distributions. It was seen in the previous chapter that data are usually put in the form of a frequency distribution. It is now proposed to show how the arithmetic mean of such a distribution can be calculated.

Example.

Age (years)	*Frequency*	ξ	$f\xi$	$f(\xi + 1)$
15–25	5	−2	−10	5(−1) = −5
25–35	17	−1	−17	17(0) = 0
35–45	37	0	0	37(1) = 37
45–55	26	+1	26	26(2) = 52
55–65	15	+2	30	15(3) = 45
	100		29	129

$$\text{Mean (denoted by } \bar{x}) = 40 + \frac{29}{100} \times 10 = 42{\cdot}9 \text{ years.}$$

An assumed average is taken. The nearer this is to the actual average the smaller the magnitude of the calculations. From the pattern of the distribution it is clear that the actual average is probably in the group with the greatest frequency. Hence the assumed average is taken as 40 years, the mid-point of the 35–45 class. Deviations are then calculated from this assumed average. These deviations are given in the column headed ξ (the Greek letter "xi"—pronounced ksee). They are in working units of 10, since this is the class interval in every case. If there had not been a common class interval, actual deviations would have to be calculated, and not given in working units, as in the example shown. The column headed $f\xi$ gives the product of the frequency and the deviation. The columns headed "frequency" and $f\xi$ are then added.

The mean is then calculated as follows:

$$\text{Assumed average} + \frac{\text{Sum of the } f\xi \text{ column}}{\text{number of items}} \times \text{working unit.}$$

Note that mathematical limits must always be used. It is assumed that the five items in the first class, 15–25, have an average value of 20, the mid-point. This assumption is justified if the grouping has been correctly done. Similar assumptions are made for the other classes.

The last column, headed $f(\xi + 1)$, is to provide an arithmetical check. For each class, 1 is added to the deviation, and this figure is then multiplied by the respective frequency. The resulting products are added. If the arithmetical calculations are correct, then the total of this column less the total of the $f\xi$ column will equal the total number of items. In the example given we have

$$129 - 29 = 100$$

thus providing a check on the arithmetic. This is known as Charlier's check.

£	Number
0–300	21
300–600	211
over 600	2
	234

Fig. 61.—A frequency distribution with open-end class.

The treatment of open-end classes. Consider the frequency distribution in Fig. 61. As will be seen, there is no upper limit to the last class. In order to compute the mean, some reasonable assumption must be made as to its magnitude. The best guide to its probable size is the pattern of the distribution. In the example given £600–£900 would probably be chosen. First, it is convenient to do so, since there would then be equal class intervals making for ease of any subsequent calculations. But more important is the pattern of distribution showing that the two items in the open-end class are not likely to be outside the limit of £900. In any case there are only two items as compared with the total of 234. Hence the amount of error will not be excessive on the assumption made.

Sales

Month	Sales (£)	Average monthly sales (£)
Jan	48	48
Feb	34	41 $\left(\text{i.e.} \frac{48+34}{2}\right)$
Mar	50	44 $\left(\text{i.e.} \frac{48+34+50}{3}\right)$
Apr	36	42 $\left(\text{i.e.} \frac{48+34+50+36}{4}\right)$

Fig. 62.—A progressive average.

The progressive average. This in effect is a recalculated average as more and more data become available (see Fig. 62). The average monthly sales in Fig. 62 is a progressive average.

The moving average. An application of the arithmetic average is the moving average.

The moving average is a simple method of indicating the trend measurement of data over a period.

To obtain the moving average, a period is taken which is long enough to eliminate the effect of ordinary seasonal or recurring fluctuations.

Consider the data in Fig. 63.

The first five years' profits are totalled and the average is taken. This figure is placed opposite the third year, *i.e.* the centre year of the five. In a 3- or 7-year average it would be placed opposite the 2nd and 4th year respectively. Next year's profit, £10,010, is included in the five years' total, and the £100 of the first year is dropped, and so on, *i.e.* one further year is included and the earliest of the previous five is dropped each succeeding year.

The harmonic average. This is the reciprocal of the average of the reciprocals of the values to be averaged.

$$\text{Harmonic average of 2, 4 and 5} = \frac{1}{\frac{1}{3}(\frac{1}{2}+\frac{1}{4}+\frac{1}{5})} = \frac{60}{19} = 3\frac{3}{19}.$$

When averaging rates such as miles per hour, income per £ of capital, etc., great care is required. Sometimes the arithmetic average is required, sometimes the harmonic. A car travels at 40 miles per hour to a certain place and back again at 30 miles per hour. The average speed is the harmonic average. A car travels for one hour at 40 miles per hour and for a further hour at 30 miles per hour. The average speed is the arithmetic average. In the first case, the distance was constant, in the second, the time.

Annual Profits of P.Q. Company

Year	Profits	Five years' moving average
1	100	–
2	200	–
3	400	600
4	700	2,582
5	1,600	4,442
6	10,010	5,562
7	9,500	6,340
8	6,000	7,020
9	4,590	6,098
10	5,000	5,218
11	5,400	5,078
12	5,100	5,280
13	5,300	5,460
14	5,600	–
15	5,900	–
	£65,400	

15 years' average = £4,360.

Fig. 63.—Series of data showing moving average.

The rule is: when rates are expressed as x per y, and x is constant, the harmonic average is required. If y is the constant element, then the arithmetic mean is required.

Questions

1. The table below shows the distribution of the miles run per vehicle for a typical day's work with a fleet of goods vehicles. Calculate the average number of miles per vehicle per day.

30–39	4	70–79	15	110–119	11	150–159	2
40–49	6	80–89	13	120–129	1		
50–59	3	90–99	8	130–139	2		
60–69	9	100–109	10	140–149	3		87

Chartered Institute of Transport.

2. A man bought 17 pints of oil at each of three separate towns. He obtained 5 pints for £1 at one town, 6 pints for £1 at the second town and 10 pints per £ at the third. What was the average number of pints he obtained for £1?

(*Note:* If the calculation is correctly done the figure of 17 should not appear in your computation.)

3. Chart the following quarterly figures for sales over a period of five years, showing a moving annual average.

	£	£	£	£
1965 . .	123	138	197	108
1966 . .	117	139	182	104
1967 . .	111	122	203	93
1968 . .	119	117	162	95
1969 . .	103	110	149	91

What is the advantage of such an average?

Institute of Cost and Management Accountants.

4. The earnings of a sample of 140 women and 200 men are given below. Calculate the arithmetic average of each group. Compare these two averages and describe the differences in the distribution of earnings for males and females.

Earnings (new pence)	*Number of* *Males*	*Females*
100 and under 105	4	12
105 " 110	18	28
110 " 115	32	36
115 " 120	46	34
120 " 125	44	16
125 " 130	42	8
130 and over	14	6
	200	140

5. Explain how weighted averages are calculated and describe the general method used to obtain the weights.

At harvesting time a farmer employed 20 women, 10 men and 16 boys to lift potatoes. The woman's work was three-quarters as effective as that of a man, while a boy's work was only half. Find the daily wage bill if a man's rate was 240p a day and the rates for the women and boys in proportion to their effectiveness. Calculate the average daily rate for the 46 workers.

6. Discuss briefly the advantages and disadvantages of the arithmetic average. Give eight cases where the arithmetic average may be used in business and use some of these to illustrate any disadvantage such an average may have.

Institute of Cost and Management Accountants.

7.

Weekly Retail Turnover of Butcher's Shops

Weekly retail turnover	*Number of*	
	Multiple shops	*Independent retailers*
£	(hundreds)	(hundreds)
Under 30	10	57
30 and under 60	26	102
60 " 90	26	67
90 " 120	16	31
120 " 150	9	15
150 " 180	6	7
180 and over	10	10
	103	289

Calculate the mean weekly retail turnover of multiple shops and independent retailers.

8. A firm buys £8,000 of plant and equipment on the 1st January 1971. Depreciation is provided for as follows:

Electrical plant . .	£5,000 (5 per cent on reducing balance).
General plant . .	£2,000 (10 per cent on reducing balance).
Vans . . .	£1,000 (20 per cent on reducing balance).

(*a*) Find the average rate of depreciation for the year 1971.
(*b*) Find the average rate of depreciation for the year 1972.
(*c*) Account for the difference between the two averages.

9. A company which makes and sells a standard article has four machines on which this article can be made. Owing to differences in age and design these machines run at different speeds, as follows:

Machine	*Number of minutes required to produce one article*
A	2
B	3
C	5
D	6

(*a*) When all machines are running what is the total number of articles produced per hour?

(*b*) Over a period of 3 hours, machines B, C and D were run for the first 2 hours, and machines A, B and D only were run for the last hour. What was the average number of articles produced *per hour* over this 3-hour period?

Institute of Cost and Management Accountants.

CHAPTER 16

THE MEDIAN AND OTHER POSITIONAL VALUES

The median. The arithmetic average is not satisfactory for all purposes. For instance, it is difficult to measure certain qualities for which an average might be required, but for which no mathematical measure can be employed. It is usually impracticable to state a definite measurement of such things as the ability of employees, advertising campaigns, endurance tests and other factors in business. Further, it is not always possible to obtain all the data necessary for the computation of the arithmetic or other calculated average. On the other hand, however, it may be possible, for instance, to arrange employees in order of competence or intelligence, or forms of advertising in order of results, and if such an array is made in order of size or importance from the lowest to the highest (or vice versa), the middle item may be regarded as the average one. In statistics this middle item of an array of data is called the median.

The median is a position average, and is the value of the mid-case in an array.

How to locate the median. All the items or numbers of a group or class are arranged in an array in order of size, the items are counted and the middle item is the required median. The median thus has the same number of items above it as below it.

The *position* of the median can be found by the formula $\frac{n+1}{2}$ in which n is the number of items in the array.

In the array arranged in Fig. 65 the median is the seventh item (*viz.* $\frac{n+1}{2} = \frac{13+1}{2} = 7$th item), which means that the median average weekly sales in three months is £103.

If the number of items in the array is *even,* the practice is to take the mean of the two middle items, as the median must lie between them. One item removed from the array in Fig. 65 would make the median the 6½th term, *i.e.* between £102 and £103, with a value of $\frac{205}{2} = £102\frac{1}{2}$.

Sales

Week no.	Weekly sales for 3 months
	£
1	102
2	106
3	98
4	100
5	103
6	99
7	102
8	108
9	106
10	104
11	99
12	109
13	104

Fig. 64.—Original data unarranged.

Sales

Weekly sales for 3 months	
£	
98	
99	
99	
100	
102	
102	
103	← median
104	
104	
107	
106	
108	
109	

Fig. 65.—Same data in a simple array in order of magnitudes.

The median of a discrete series. When the items have been grouped by sizes (see Fig. 66), the numbers occurring for each size are accumulated in a cumulative frequency column. The position, or rank, of the median is then $\frac{n+1}{2}$. In the example given this will be $\frac{4{,}609+1}{2} = 2{,}305$. By inspecting the cumulative frequency column, it will be seen that the value of this item is $8\frac{1}{2}$. This is the median size of shoes sold. The series in this example is a non-continuous or discrete series. The sizes of the variable, the sizes of the shoes, are of definite amounts. The shoes are made only in the sizes indicated.

The median of a grouped distribution. In the case of a grouped distribution, each class covers a number of sizes, varying from a lower to an upper limit. Whether the sizes of the individual items are continuous or discrete, when they are grouped it may be assumed that the items in each class are equally distributed in order of size within each class, and that the average value of the items in each group is equal in value to the mid-point value of the class interval. The rank of the median in a grouped distribution is, therefore, $\frac{n}{2}$. The value of this item is found by interpolation.

Fig. 67 shows a grouped distribution. The rank of the median is $\frac{301}{2} = 150\frac{1}{2}$. It is required to find the value of this item.

X Co.: Numbers of Shoes Sold by Sizes in One Year

Size	Number of pairs	Cumulative total
5	30	30
5½	40	70
6	50	120
6½	150	270
7	300	570
7½	600	1,170
8	950	2,120
M=8½	820 median in this total	2,940
9	750	3,690
9½	440	4,130
10	250	4,380
10½	150	4,530
11	40	4,570
11½	39	4,609
	Total 4,609	

Fig. 66.—A frequency distribution which is discrete or non-continuous.

It will be seen that the median is in the class £2·00–£2·50. The value of the 75th item is £2·00. The rank of the median is 150½. The median is therefore 150½ − 75 items, *i.e.* 75½ items greater than £2·00. In the class containing the median there are 150 items, assumed to be equally distributed between £2·00 and £2·50. 150 items increase regularly until the last item is 50p greater than the first; therefore the 75½th item will have increased by $\frac{75\frac{1}{2}}{150} \times 50\text{p} = 25\text{p}$. The median is therefore £2·00 + 25p = £2·25.

Finding the median graphically. A cumulative frequency curve is drawn as described in Chapter 14. The median position on the

Number of Employees Engaged at Various Daily Wage Rates

Wage groups	No. of employees	Cumulative total
£ £		
1·00 and under 1·50	30	30
1·50 and under 2·00	45	75
2·0 and under 2·50	150	225
2·50 and under 3·00	36	261
3·00 and under 3·50	24	285
3·50 and over	16	301
	Total 301	

Fig. 67.—A grouped distribution.

vertical axis will have a corresponding value on the horizontal axis. This value will be the value of the median. This is shown in Fig. 68. A dotted line is drawn from the "vertical" axis from point 150½ horizontally until it reaches the curve. At this point a perpendicular

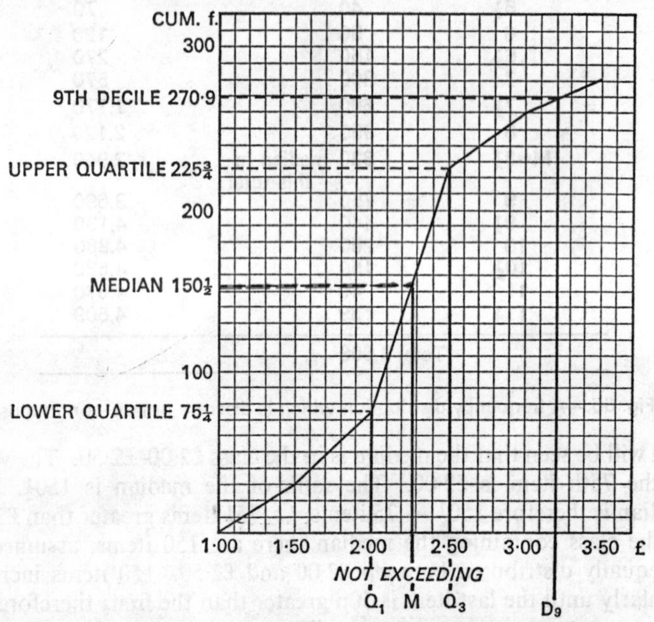

Fig. 68.—Graphical method of finding median, quartiles, etc.

line is dropped on to the x-axis, which then gives the value of the median.

If the points plotted on a cumulative frequency curve are joined by straight lines, the graphical result will agree with the arithmetical result.

Advantages and disadvantages of the median.

1. *Advantages.*

(*a*) If found directly, it is representative of an actual item.

(*b*) It is simple to understand.

(*c*) Extreme items do not affect its value.

(*d*) It can be obtained even when the values of all items are not known. Provided the middle items are known, and that there are the same number of larger and the same number of smaller items, the median can be located.

(*e*) It can be used for measuring qualities and factors to which mathematical measurement cannot be given.

2. *Disadvantages.*

(*a*) If the items are few, it is not likely to be representative.

(*b*) If the distribution is irregular, the location of the median may be indefinite.

(*c*) The arranging of data in the necessary array is often tedious.

(*d*) It cannot be used to determine the total value of all the cases or items. The number of items multiplied by the median will not give the total of the data. It is not suitable for arithmetical calculations, and has but limited use in practical work.

Quartiles, deciles and percentiles. In the same way as the median divides the array into two parts, one part having half the values greater than the median, the other having half the values lower than the median, so the quartiles divide the array into four parts, the deciles divide it into ten parts and the percentiles divide it into one hundred parts.

The lower quartile will have one-quarter of the values lower and three-quarters of the values higher than the value of the lower quartile. The upper quartile will have one-quarter of the values higher and three-quarters of the values lower than the value of the upper quartile. The second quartile is, of course, the median.

The first decile would have one-tenth of the values lower than the value of the first decile and nine-tenths of the values higher. The second decile would have two-tenths of the values lower and eight-tenths of the values higher than the value of the second decile, and so on.

The 35th percentile would have 35-hundredths of the values lower than the value of the 35th percentile and 65-hundredths of the values higher. Similarly for all other percentiles.

Evaluation of quartiles, deciles and percentiles. The method of calculation is exactly the same as for the median. The arithmetical method will require a cumulative frequency table and the graphical method a cumulative frequency curve (see Fig. 68).

$$Q_1 \text{ Rank} = \frac{301}{4} = 75\tfrac{1}{4}.$$

Value of 75th item = £2·00. $\frac{1}{4}$ item increases value by $\frac{\frac{1}{4}}{150} \times$ 50p.

$$\text{Value of } Q_1 = £2{\cdot}00 + {\cdot}001 = £2{\cdot}001.$$

$$Q_3 \text{ Rank} = \frac{3 \times 301}{4} = 225\tfrac{3}{4}.$$

Value of 225th item = £2·50. $\frac{3}{4}$ item increases value by $\frac{\frac{3}{4}}{36} \times$ 50p.

	Rank for discrete series	Rank for grouped distribution
Lower quartile	$\frac{n+1}{4}$	$\frac{n}{4}$
Upper quartile	$\frac{3(n+1)}{4}$	$\frac{3n}{4}$
Ninth decile	$\frac{9(n+1)}{10}$	$\frac{9n}{10}$
Twenty-seventh percentile	$\frac{27(n+1)}{100}$	$\frac{27n}{100}$

Fig. 69.—Ranks of quartiles, deciles and percentiles.

$$\text{Value of } Q_3 = £2{\cdot}50 + {\cdot}01 = \underline{£2{\cdot}51}$$

$$\text{Ninth decile. Rank} = \frac{9 \times 301}{10} = 270{\cdot}9.$$

$$\text{Value of ninth decile} = £3{\cdot}00 + \frac{9{\cdot}9}{24} \times 50\text{p} = £3{\cdot}20.$$

Questions

1. The following table shows the frequency distribution of actual headways between buses which were observed at one point on an in-town route with a scheduled frequency of 30 buses per hour. Calculate (*a*) the arithmetic average headway, (*b*) the median headway and (*c*) the two limits—one above and one below the median—between which it is estimated that half the number of headway observations occurred.

Headway (nearest minute)	*Number of buses*
0	111
1	189
2	117
3	75
4	48
5	30
6	15
7	8
8	3
9	3
10	1

Chartered Institute of Transport.

2. Using the data given below find the median and quartile number of employees per firm. Use graphical methods.

Number of employees	*Number of firms*
Under 10	41
10–20	96
20–30	143
30–40	85
40–50	36
50–60	11
60–70	7
70–80	4
80–90	3
90–100	1

Institute of Company Accountants.

3.

Cinemas in Great Britain

Size of cinema (seats)	*Number of cinemas*
Up to 250	165
251–500	897
501–750	1,043
751–1,000	895
1,001–1,250	589
1,251–1,500	334
1,501–1,750	264
1,751–2,000	206
Over 2,000	204

Using graphical methods estimate the median and quartile size of cinema.

Institute of Company Accountants.

4. Comment on the suitability of the class intervals below for use in connection with data, of which a sample item would be 29·9.

A	*B*	*C*
0–10	0–9·9	0–9·95
10–20	10–19·9	9·95–19·95
20–30	20–29·9	19·95–29·95
30–40	30–39·9	29·95–39·95

5. State briefly and concisely the chief fallacies in the use of averages.

6. A survey of all the firms within an industry revealed the following:

Turnover p.a. £(000s)	*Number of firms*
below 50	5
50–100	8
100–150	9
150–200	12
200–250	18
250–300	23
300–350	17
350–400	14
400–450	5
450–500	1

Calculate the mean value of sales having assumed the mean is £200,000. Comment on the accuracy of the resultant mean value.

Association of Certified Accountants.

7. Arrange the following data in a frequency distribution. From the grouped distribution calculate the median and the arithmetic mean. Compare these values with the corresponding values computed from the original data. 140p; 150p; 180p (4 employees); 190p (2 employees); 200p (7 employees); 205p; 210p (4 employees); 220p (4 employees); 230p (2 employees); 240p (8 employees); 250p (7 employees); 255p; 270p; 280p; 290p; 300p (10 employees); 310p; 320p; 325p; 350p; 360p; 380p; 400p (3 employees); 450p (6 employees); 495p and 550p. (Total employees = 72.)

CHAPTER 17

THE MODE AND THE GEOMETRIC MEAN

The mode. The mode is what many people mean in ordinary conversation when they refer to such things as an average income, an average person, etc. An example which is most frequent or typical is implied. It is the term used to designate the most frequent item in a series, the value which occurs most in a group, or the value of greatest density. The mode may therefore be defined as the size of the variable that occurs most frequently, or the value about which most cases recur. The mode is the value of the maximum ordinate in a smoothed histogram.

Like the median, the mode has very limited practical use, and cannot be used for arithmetic manipulation. It is of value, however, for the purposes of, say, the maker of ready-made clothes, or the manufacturer of certain components or accessories in common use. In deciding upon wages and rates of pay, an employer will usually adopt the modal rate—the rate paid by most other employers. Many other practical points in business are dependent upon some modal feature, particularly in regard to wages, rents, prices, sizes of supplies and the like. Not infrequently modal information on many of these matters can be supplied straight away by persons of experience. On the other hand, the mode may not be well defined, and in some cases the "point of density" may be rather broad and the selection of the mode may be based merely on judgment or even desire.

In a frequency distribution, it is the point of greatest frequency of occurrence.

Like the median, it is a position average, and for practical computation it is usually essential to arrange the data in a frequency distribution or array. If such a distribution arrangement does not show a well-defined mode or modal class the modal average should not be used.

How to find the mode (by formula). The following formula is useful to find the mode when there is a regular class interval.

The formula is:

Let $Z =$ mode,

l_1 and $l_2 =$ the lower and upper limits of the modal group,

$f_1 =$ the number of items or frequency in the modal group,

f_0 and f_2 = the frequencies in the next lower and next higher groups.

Then $Z = l_1 + \dfrac{f_1 - f_0}{2f_1 - f_0 - f_2}(l_2 - l_1)$.

(*Note:* Sometimes c is used to represent $(l_2 - l_1)$.)

Substituting the values in Fig. 57, we have

$$Z = £2{\cdot}00 + \frac{150 - 45}{300 - 45 - 36} \times (2{\cdot}50 - 2{\cdot}00) = £2{\cdot}00 + \frac{105}{219} \times {\cdot}50$$

$= £2{\cdot}24$ which is the modal wage.

Finding the mode graphically. The method shown below (Fig. 70) can be used only when a grouped distribution is available. The class intervals of the modal group and the two groups either side must be equal. The result will be the same as the arithmetical method just given.

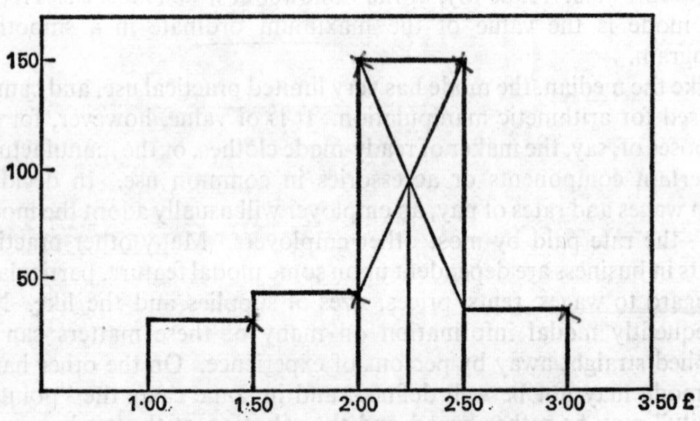

Fig. 70.—Graphical method of finding the mode.

Advantages and disadvantages of the mode.

1. *Advantages.*
   (*a*) It is easy to understand.
   (*b*) Extreme items do not affect its value.
   (*c*) Like the median, only the middle items need be known.

2. *Disadvantages.*
   (*a*) It is often not clearly defined.
   (*b*) Exact location is often uncertain.
   (*c*) Arrangement of data is tedious.

Series with more than one mode. A series may have more than one mode, and is then said to be bi-modal, tri-modal, etc. In practice, there is usually one prominent mode, with sub-modes, although several modes may be of equal importance. This might happen with data relating to the sale of goods, there being different modes for low-, medium- and high-priced articles. By separating the data for each grade, a uni-modal series for each of them is obtainable.

The geometric mean. The geometric mean is obtained by taking the nth root of the product of "n" values. This is best explained by means of an example.

Example.

Find the geometric mean of 1,421 ; 1,082; 1,092. Here there are three items; the G.M. (geometric mean) will be the cube (*i.e.* third) root of the product of these three numbers.

$$\text{G.M.} = \sqrt[3]{1{,}421 \times 1{,}082 \times 1{,}092}$$

Number	*Logarithm*
1,421	3·1526
1,082	3·0342
1,092	3·0382
	3)9·2250
1,189	3·0750

G.M. = 1,189.

Advantages and disadvantages of the geometric mean.

1. *Advantages.*

(*a*) It makes use of all the data in a group, and can be calculated with mathematical exactness, provided that all the quantities are greater than zero and positive.

(*b*) It is the only average that can be used to indicate *rate* of change, *e.g.* from 1968 to 1971 prices increased 5 per cent, 10 per cent and 18 per cent respectively.

What was the average annual increase? It was not 11·0 per cent as shown by the arithmetic average $\frac{5+10+18}{3} = 11$, but 10·9 per cent, as shown by the geometric mean, *viz.* $\sqrt[3]{1{\cdot}05 \times 1{\cdot}10 \times 1{\cdot}18} =$ 1·109, *i.e.* a 10·9 per cent increase.

(*c*) Large items have less effect on it than in the arithmetic average.

2. *Disadvantages.*

(*a*) It is impossible to use it when any items are zero or negative.

(*b*) It is more difficult to calculate and less easily understood.

(*c*) It may locate a value which does not correspond with any actual example.

Questions

1. A man get three annual rises in salary. At the end of the first year he gets an increase of 4 per cent, at the end of the second an increase of 6 per cent on his salary as it was at the end of the year, and at the end of the third year an increase of 9 per cent on his salary as it was at the end of the third year. What is the average percentage increase?

2. Define a "weighted average."

Dept.	*Number of workmen*	*Total wages*
A	432	£3,623
B	517	6,837
C	117	1,152

A cost-of-living bonus is given to each workman amounting to £2·00. What is the average percentage increase per man for each department and for the total?

3. When is the mode a suitable average to use? State its disadvantages.

4. What is the median? What are its advantages and disadvantages?

5. From the following table showing the wage distribution in a certain factory determine:

(*a*) the mean wage,
(*b*) the median wage,
(*c*) the modal wage,
(*d*) the wage limits for the middle 50 per cent of the wage earners,
(*e*) the percentage of workers who earned between £7 and £10,
(*f*) the percentage who earned more than £6 per week,
(*g*) the percentage who earned less than £5 per week.

Weekly wage	*Number of employees*
£	
3·5–4·5	17
4·5–5·5	23
5·5–6·5	42
6·5–7·5	52
7·5–8·5	78
8·5–9·5	117
9·5–10·5	223
10·5–11·5	82
11·5–12·5	14
12·5–13·5	3

6. Write brief notes on the following topics:

(*a*) secondary statistics,
(*b*) mode,
(*c*) ratio scales,
(*d*) the median.

7. From the following frequency distribution of the hourly earnings of the skilled employees of Electronics Ltd. (Workshop I):

(*a*) Calculate the mean hourly wage and the modal hourly wage, and
(*b*) Explain the characteristics of the mean and mode as representative values.

Electronics Ltd. (Workshop I)

Hourly wage (p)	*Number of employees*
50–59·9	8
60–69·9	10
70–79·9	16
80–89·9	14
90–99·9	10
100–109·9	5
110–119·9	2

Association of Certified Accountants.

CHAPTER 18

INDEX NUMBERS

The use of index numbers. It is useful for many purposes to know how the price level of a group of commodities has changed over time. A price index enables this to be done. It is an average of the changes in the price of the individual items in the group. Similarly a quantity or volume index shows the average changes in the quantities of the items of a group of commodities. Examples of price indices include the retail price index, the index of wholesale prices, the index of weekly wage rates (wages being the price of labour) and the index of agricultural prices. Quantity indexes include the index of industrial production, and the indices of the volume of exports and imports. Descriptions of many official index numbers are given in Chapter 28.

Index numbers are useful for showing trends; they enable comparisons to be made between movements in the levels of prices of different groups of commodities, or between the levels of prices and wages, or between the levels of production and wages, and so on.

Index numbers can also be used to make comparisons over space. It is also possible to construct index numbers to show movements in productivity within a firm. *An index number is a device which shows by its variations the changes in a magnitude which is not capable of accurate measurement in itself or of direct valuation in practice*. It is an indirect measure of a concept, *e.g.* changes in price level. As such, there are various methods of compiling an index number. Naturally different methods give different results. Since, however, only a series of index numbers has any meaning, provided the same method of compilation is used throughout the series, this is not important.

Compilation of index numbers. It is now proposed to show the usual method of compilation of an index number. Most of the other methods are of academic interest only, and will not be dealt with.

In order to illustrate the method and to keep calculations to a minimum, only three commodities have been chosen. A price index and a quantity index will be computed.

The data are as follows:

Item	Unit	3rd August 1968 Quantity	3rd August 1968 Price	9th November 1971 Quantity	9th November 1971 Price
			p		p
Bread	per loaf	3	8	6	12
Butter	per lb	1	45	1½	36
Cheese	per lb	½	30	1	36

The quantities are the average weekly purchases of these commodities of an imaginary household for the dates given. The prices are those ruling on the dates given. The price index number for 9th November 1971 will show the price level of these commodities as a percentage of the price level at 3rd August 1968. The quantity index for 9th November 1971 will show the average weekly quantity bought for this date as a percentage of the average weekly quantity bought for 3rd August 1968.

It will subsequently be shown that the price index is 107, which means that prices have risen by 7 per cent between these two dates. It will also be shown that the quantity index is 173, which means that the average weekly quantity bought has increased by 73 per cent.

The date with which comparison is made is known as the base date, in this case 3rd August 1968.

The price index will be computed first. The first step is to calculate the *price relatives.* A price relative is the price at the current date expressed as a percentage of the price at the base date. In the example given the price relatives are as follows:

$$\text{Bread } \frac{12\text{p}}{8\text{p}} \times 100 = 150.$$

$$\text{Butter } \frac{36\text{p}}{45\text{p}} \times 100 = 80.$$

$$\text{Cheese } \frac{36\text{p}}{30\text{p}} \times 100 = 120.$$

To obtain the price index, a weighted average of these price relatives is calculated. The importance of these various commodities, and therefore the weights, will be proportional to the expenditure on each item. The expenditure will be that of the base date. The calculation is shown below.

Item	Expenditure at base date	Price relative	Price relative × weight
	p		
Bread	24	150	3,600
Butter	45	80	3,600
Cheese	15	120	1,800
	84		9,000

$$\text{Price Index} = \frac{9{,}000}{84} = 107.$$

Index numbers are calculated to the nearest whole number, unless they are to be used for further calculations.

The quantity index is computed in a similar way. It will be a weighted average of quantity relatives, the weights being the same as for a price index. The calculations are as follows:

Item	*Quantity relatives*	*Weight*	*Quantity relative × weight*
Bread	$\frac{6 \text{ loaves}}{3 \text{ loaves}} \times 100$	24	$200 \times 24 = 4{,}800$
Butter	$\frac{1\frac{1}{2} \text{ lb}}{1 \text{ lb}} \times 100$	45	$150 \times 45 = 6{,}750$
Cheese	$\frac{1 \text{ lb}}{\frac{1}{2} \text{ lb}} \times 100$	15	$200 \times 15 = 3{,}000$
			14,550

$$\text{Quantity Index} = \frac{14{,}550}{84} = 173.$$

The above method of calculation is known as the average of ratios method. *The weights are values.*

The aggregative method. This will give the same results as before, provided that base-date quantities and prices are used as weights. The price index will be the value of base-date quantities at current prices, expressed as a percentage of base-date quantities at base-date prices. It expresses the value of a group of commodities at current prices as a percentage of the value of the same group of commodities at base-date prices. The quantity index will be the value of the current quantities at base-date prices expressed as a percentage of the value of the base-date quantities at base-date prices. It expresses the value of current quantities at base-date prices as a percentage of the value of base-date quantities at the same prices.

Using the same data as before, the price index is calculated as follows:

Item	*Value at date*	*Value of base-date quantities at current prices*
	p	p
Bread	24	3 loaves at 12p = 36
Butter	45	1 lb at 36p = 36
Cheese	15	½ lb at 36p = 18
	84	90

$$\text{Price Index} = \frac{90\text{p}}{84\text{p}} \times 100 = 107 \text{ (as before).}$$

The quantity index is calculated as follows:

Item	*Value at base date*	*Value of current quantities at base-date prices*
	p	p
Bread	24	6 loaves at 8p = 48
Butter	45	$1\frac{1}{2}$ lb at 45p = $67\frac{1}{2}$
Cheese	15	1 lb at 30p = 30
	84	$145\frac{1}{2}$

$$\text{Quantity Index} = \frac{145\frac{1}{2}\text{p}}{84\text{p}} \times 100 = 173 \text{ (as before).}$$

A theoretical objection to the methods just given is that the price index multiplied by the quantity index will not indicate changes in value. If, however, in calculating the price index, current quantities had been taken instead of base-date quantities, the quantity index being calculated with base-date prices, the product would have shown the changes in value. There is, however, the very serious practical difficulty that when compiling price indices, it is often not possible to obtain current quantities. In practice, the theoretical objection is not important.

The chain base method. Circumstances may arise in which it is possible to compare the prices of one year with the next, but it is not possible to compare prices when separated by a large number of years. If new commodities are replacing old ones, or the weights are rapidly changing, the methods just described will not be possible. In this case, it may be useful to compile indices with the previous year as base and to "chain" them so as to get a series referring back to a base year. Movements between successive years will be as accurate as those compiled by the methods just described, but for movements between years far apart, the series may be very inaccurate.

Example. Price index for 1969 (1968 = 100) = 110.
Price index for 1970 (1969 = 100) = 105.
Price index for 1971 (1970 = 100) = 125.

Chain indices are obtained as follows:

$$\text{Price index for 1970 (1968 = 100)} = \frac{110 \times 105}{100} = 115{\cdot}5.$$

$$\text{Price index for 1971 (1968 = 100)} = \frac{110 \times 105 \times 125}{100 \times 100} = 144.$$

The series (1968 = 100) is 100; 110; 116; 144.

Problems of index number construction. The first essential point to be considered is the object for which the index number is to be constructed. What is it to measure and why? Suppose, for example, it is required to measure the amount of production to show the trend of economic activity. Suppose, further, this is required monthly. The data which can be made available will determine the scope of the index. Since it is to be monthly, agricultural production will have to be omitted. It will also be difficult to include many forms of production. The problem of the scope of the index is bound up with the purpose of the index and the data available. The data available, or rather the lack of it, may necessitate the modification of the purpose. Instead of an index of production as a whole being sought, an index of industrial production might now become the aim.

The next problem is the selection of items. Consider the compilation of a price index. Usually, it is not practicable, either from the point of view of cost or time, even if it is possible, to measure changes in the prices of all the relevant commodities. Hence a selection must be made. These must be selected so that movements in the prices of those chosen will be representative of the movements of prices of all the relevant commodities.

The choice of the base date or period presents some difficulties. However, there is a general misconception that a "normal" year or date must be chosen, otherwise the weights soon become incorrect. In the case of a quantity index the base date or period must only be such that the prices of the various commodities are reasonably related, and in the case of a price index, a base date or period when the quantities of the commodities are reasonably related should be chosen.

The choice of weights provides another set of problems. Where fewer items are chosen as indicators than all those relevant, the weight must be that relating to the whole of those relevant. Thus, in the case of the index of retail prices, only five items are taken to represent floor coverings, but the weight is in respect of floor coverings as a whole. In the case of the index of production, the weights used are the values of net outputs of the various industries. Sometimes difficulty arises because the value of the net output of an industry has to be apportioned over a number of products.

The form of average is usually dictated by practical considerations. Thus, despite the theoretical objections to the methods, described earlier in the chapter, they are the most usual.

Limitations of index numbers. The index is usually based on a sample, hence sampling errors are introduced. It is not possible to take into account all changes in quality or product. Comparisons over long periods are not reliable. Different methods of compilation give different results, but, unless there are rapid changes in conditions, the trends generally agree.

Index numbers formulae.

p_0 represents the price at the base date.
p_1 represents the price at the current date.
q_0 represents the quantity at the base date.
q_1 represents the quantity at the current date.

1. *Price Index.*

$$\text{Average of ratios} \quad \frac{\Sigma \frac{p_1}{p_0}(p_0q_0)}{\Sigma p_0q_0}.$$

$$\text{Aggregative method} \quad \frac{\Sigma p_1q_0}{\Sigma p_0q_0}.$$

2. *Quantity Index.*

$$\text{Average of ratios} \quad \frac{\Sigma \frac{q_1}{q_0}(p_0q_0)}{\Sigma p_0q_0}.$$

$$\text{Aggregative method} \quad \frac{\Sigma q_1p_0}{\Sigma q_0p_0}.$$

Both these index numbers are *Laspeyres* indices; they have base-period quantities and base-date prices for weights. Index numbers which have current-period quantities and current-date prices are called *Paasche* indices.

Questions

1. It has been stated that the technique of index number construction involves four major factors:

(*a*) Choice of items.
(*b*) Base period.
(*c*) Form of average.
(*d*) Weighting system.

Do you agree with this view? If so, explain these four factors and discuss the problems to which they give rise. If you do not agree, give your views on the main problems involved in index construction.

Chartered Institute of Transport.

2. In the manufacture of a particular product, two commodities, *X* and *Y*, are considered to be equally important in 1971. Their prices in June 1971 and 1972 were as follows:

	1971	*1972*
X £ per cwt . .	8	12
Y pence per foot . .	30	20

Show the unweighted price relatives for each commodity using 1971 and 1972 as bases, and comment on the results.

How would a combined index for *X* and *Y* be worked out?

3. Average prices in shillings per quarter.

	1950	*1951*	*1952*
Wheat . . .	110·7	122·8	129·0
Barley . . .	99·7	124·5	116·3
Oats . . .	60·1	72·9	74·4

Take the price averages in 1950 as 100 and construct a price index number for the three cereals together, for 1951 and for 1952 using the weights 8 for wheat, 5 for barley and 3 for oats.

4.

Barbalonia Imports of Dutiable Beverages

	1960		1971	
	Quantities	*Value* (£)	*Quantities*	*Value* (£)
Beer and Ale (barrel)	58,577	158,650	448	4,057
Cocoa, raw (lb) .	5[illegible]670,321	1,335,107	246,623,216	8,943,025
Coffee (cwt) .	765,561	2,024,648	1,066,046	5,988,812
Tea (lb) . .	321,190,064	9,904,085	494,353,466	33,050,853
Wine (gallon) .	13,103,304	4,214,878	25,252,387	18,167,077

Compute (*a*) a volume index for 1971, using 1960 as the base date; (*b*) a price index for 1960, using 1971 as base date.

(*Note:* Average price for each commodity is found by dividing values by quantities. Hence price relatives can be found.)

5. The following table indicates the price per ton, and value of sales for a raw material:

Grade of material	1968 *Price* (£)	*Sales* (£ *mn.*)	1974 *Price* (£)	*Sales* (£ *mn.*)
A	6·50	13	8·75	26·25
B	9·00	81	14·00	140·00
C	8·50	34	10·00	50·00

Calculate a Laspeyres Index for 1974 (all grades) taking 1968 as 100.

I.C.S.A. Part 1, June 1975.

6. A steel stockist notices that prices and values of sales for the main units of steel supply were:

Type of steel items	1969 *Price per tonne* (£)	*Sales* (£ *mn.*)	1974 *Price per tonne* (£)	*Sales* (£ *mn.*)
Ingots	162	324	200	600
Steel bars	188	564	190	760
Steel strip	220	880	275	1,100

Calculate a Paasche Index number for 1974 prices, taking 1969 = 100.

I.C.S.A. Part 1, Dec. 1975.

CHAPTER 19

DISPERSION

Two groups of numbers may easily have the same average; in one group, however, the individual items may have values near the value of the average, whereas in the other the values of the individual items may vary in value considerably from the average. The average measures the value around which the values of the individual items vary. To have any knowledge of the group, it is necessary to measure the extent to which the values of the individual items vary around the average. Such measures are known as measures of dispersion.

The range. This measure of dispersion is the difference between the highest and lowest value of the variable. It does not take into account any values except the extreme values. It is of little practical importance, except for its use in Quality Control (see Chapter 26).

The mean deviation. The sum of the deviations from the arithmetic mean is zero; hence it is necessary to ignore the sign of the deviations in computing this measure of dispersion.

Example 1.

Find the mean deviation of 3, 5, 8, 11 and 13.
The arithmetical mean is 8.
Deviations from mean are: −5, −3, 0, +3 and +5.
The total of these deviations is, of course, zero.
Ignoring the signs, the total is 16.
Mean deviation is 16 divided by 5 = 3·2.

Deviations may also be taken from the median.

The quartile deviation. The formula for this measure of dispersion is as follows:

$$\text{Q.D.} = \tfrac{1}{2}(Q_3 - Q_1).$$

This measure of dispersion, also known as the semi-interquartile range, is not a measure of deviation from any particular average.

The standard deviation. This measure of dispersion is the most important, and particular attention must be paid to the method of computation. Standard deviation is sometimes indicated by the initials S.D. and sometimes by the Greek letter sigma σ. The formula is as follows:

$$\text{S.D.} = \sqrt{\frac{\Sigma fd^2}{N}}.$$

N is the total number of items; d refers to deviations from the arithmetic mean; these are squared and then multiplied by the frequency; these products are summed; this total is indicated in symbols by Σfd^2; this sum is then divided by the number of items; the square root is then calculated; this is the standard deviation.

Example 2.

x	*f*	*d*	*fd*	*fd*²
4	2	−3	−6	18
5	3	−2	−6	12
6	2	−1	−2	2
21	1	+14	+14	196
	8			228

Notes: $fd^2 = fd \times d$,
e.g. $18 = -6 \times -3$.
Since deviations are from mean $\Sigma fd = 0$.
Mean = $(4 \times 2 + 5 \times 3 + 6 \times 2 + 21 \times 1) \div 8 = 7$.

$$\text{S.D.} = \sqrt{\frac{228}{8}} = 5{\cdot}3.$$

Short-cut method of calculating S.D. To avoid unnecessary calculations, deviations are calculated from an assumed mean, but a more complicated formula is then required. It is as follows:

$$\text{S.D.} = \sqrt{\frac{\Sigma f\xi^2}{n} - \left(\frac{\Sigma f\xi}{n}\right)^2}.$$

ξ are the deviations from any convenient assumed average. Note the difference between the real and the assumed average is $\frac{\Sigma f\xi}{n}$.

When the class interval is used as a working unit, do not forget to multiply by the value of this working unit.

Example 3.

x (*inches*)		ξ	$f\xi$	$f\xi^2$
3–6	1	−1	−1	1
6–9	4	0	0	0
9–12	3	+1	3	3
12–15	1	+2	2	4
	9		4	8

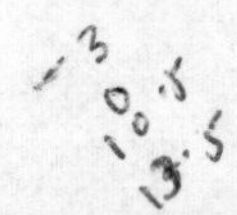

The assumed average is 7·5.
$\Sigma f\xi = 4$; $\Sigma f\xi^2 = 8$.
Working unit = 3; $n = 9$.
Substituting these values in the formula,

$$\text{S.D.} = 3 \times \sqrt{\frac{8}{9} - \left(\frac{4}{9}\right)^2} = 2{\cdot}5 \text{ inches.}$$

Relative measures of dispersion. The measures so far dealt with measure the amount of dispersion in units of years, inches, cwt, etc. In order, however, to compare variations of different groups, *e.g.* whether heights vary more or less than weights, or whether the sizes of one kind of shop vary more or less than some other kind of shop, or if the heights of girls in Glasgow vary more than the heights of girls in Aberdeen, coefficients of deviation must be compared. The relation between the measures of dispersion already given and the relative measures of dispersion or deviation are as shown in the following table.

$$\text{Coefficient of mean deviation} = \frac{\text{mean deviation}}{\text{mean or median}}.$$

$$\text{Coefficient of quartile deviation} = \frac{Q_3 - Q_1}{Q_3 + Q_1}.$$

$$\text{Coefficient of standard deviation} = \frac{\text{S.D.}}{\text{mean}}.$$

The coefficient of variation. This is the coefficient of standard deviation, just given, multiplied by 100, that is it expresses the S.D. as a percentage of the mean.

In Example 3, the mean is 7·5 + 1·3 = 8·8; the C.V. = 2·5 divided by 8·8 and then multiplied by 100, *i.e.* 28·4 per cent.

Variance. This is the square of the standard deviation, in symbols denoted by σ^2.

Questions

1. From the following figures say which class of grocers vary more in size assuming that the value of sales is the measure of size.

Value of sales	*Grocers with meat*	*Grocers with off-licence*
Under £1,000	29	128
£1,000–£2,499	189	766
£2,500–£4,999	626	2,130
£5,000–£9,999	673	2,437
£10,000–£24,999	761	2,255
£25,000–£49,999	286	772
£50,000–£99,999	74	170
£100,000–£250,000 } £250,000 and over }	7	45
	2,645	8,703

Source: *Census of Distribution and other services, 1950. H.M.S.O.*
(*Note:* It is necessary to compare the C.V.) *Institute of Statisticians.*

2. Write brief notes upon the following statistical terms:

(*a*) Weighted average (*b*) Harmonic average
(*c*) Standard deviation (*d*) Coefficient of Variation

Chartered Institute of Transport.

3. The following series of data gives the times taken in minutes by clerks to fill up a case record summary.

Calculate:

(*a*) the mean time taken to complete a summary;
(*b*) the standard deviation of the time.

From your results state what you think would be a reasonable maximum time to allow for the completion of one summary.

Number of minutes	1	2	3	4	5	6	7	8	9	10	
Number of clerks	2	3	5	10	15	30	25	15	10	5	Total: 120

Institute of Statisticians.

4. The following figures show the average number of minutes by which certain passenger trains ran late on weekdays:

	Four weeks ending:		
	27th Dec. 1967	*25th Dec. 1968*	*31st Dec. 1969*
Diesel	2·66	3·42	2·38
Electric	2·74	4·11	1·87

You are called upon to interpret these figures for the benefit of your management. Give your interpretation and explain the uses of a measurement of dispersion in this connection. *Chartered Institute of Transport.*

5. From the following figures abstracted from the Monthly Digest of Statistics calculate whether the ages of females in Northern Ireland vary more than the ages of males.

Estimated Age Distribution of Population of Northern Ireland at 30th June 1949
(Thousands)

Age group	*Males*	*Females*
0–4	72	69
5–9	66	63
10–14	56	54
15–19	57	55
20–24	61	59
25–29	57	57
30–34	48	52
35–39	43	47
40–44	38	44
45–49	33	39
50–54	29	34
55–59	28	31
60–64	27	29
65–69	24	25
70 and over	33	41

Institute of Statisticians.

6. What is the average age (to nearest 3 months) of the following fleet of double-decked public service vehicles at mid-1971?

Year of purchase	*Number*
1950	1
1951	3
1952	—
1953	3
1954	—
1955	23
1956	16
1957	1
1958	25
1959	9
1968	20
1969	8
1970	16

What measure of dispersion would you recommend for purposes of describing the scatter of age about the mean age? What is the value of this statistic (to the nearest 3 months)? *Chartered Institute of Transport.*

7. Calculate the arithmetic mean and the standard deviation of the following distribution, stating any necessary assumptions.

Heights of National Service Men

Heights (inches)	*Number of men (hundreds)*
Under 62	6
62 and under 63	10
63 ,, 64	21
64 ,, 65	39
65 ,, 66	60
66 ,, 67	80
67 ,, 68	91
68 ,, 69	88
69 ,, 70	71
70 ,, 71	53
71 ,, 72	33
72 ,, 73	19
73 ,, 74	9
74 and over	6
	586

Institute of Statisticians.

8.

Age Distribution of Women, Barbalonia, 1975

Years of age	*Millions*
Under 10	3·7
10 and under 20	3·3
20 " 30	3·7
30 " 40	3·9
40 " 50	3·7
50 " 60	3·1
60 " 70	2·4
70 " 80	1·4
80 and over	0·4

Calculate the standard deviation and coefficient of dispersion for the above distribution. Mention any other method of calculating dispersion with which you are acquainted.

9. (*a*) Distinguish between absolute and relative dispersion.

(*b*) In a final examination in statistics the mean mark of a group of 150 students was 78 and the standard deviation 8; in accountancy the mean mark of the group was 73 and the standard deviation was 7·6.

In which subject was the relative dispersion greater?

Association of Certified Accountants.

CHAPTER 20

SKEWNESS

The nature of skewness. A histogram or a frequency curve will show from its shape if a distribution is symmetrical or asymmetrical. In this chapter a number of methods will be given for measuring the amount of asymmetry or skewness. When the distribution is symmetrical, the mode, median and the arithmetical mean are all equal. The effect of skewness is to make these unequal, the value of the median being between the values of the mode and the arithmetical mean. If the mean is greater than the mode, the skewness is said to be positive; if the mode is greater than the mean, the skewness is said to be negative. When the skewness is positive, the larger values vary more than the lower, the tail is longer on the right; the distribution is said to be skewed towards the right. Figs. 71, 72 and 73 show frequency curves which are symmetrical, positively skewed and negatively skewed. A distribution which is positively skewed means that the majority of values are less than the mean, *e.g.* incomes. A distribution which is negatively skewed means that the majority of values are greater than the mean, *e.g.* age of death.

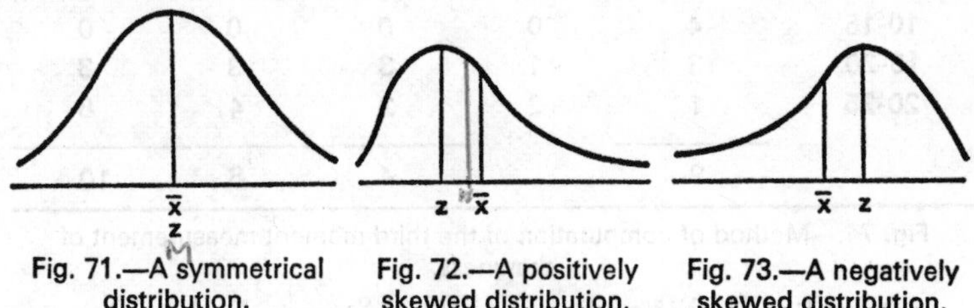

Fig. 71.—A symmetrical distribution.

Fig. 72.—A positively skewed distribution.

Fig. 73.—A negatively skewed distribution.

The Pearsonian measure of skewness. The formula is as follows:

$$\text{sk.} = \frac{\text{mean} - \text{mode}}{\text{S.D.}}.$$

When the distribution is symmetrical, since the value of the mode and the mean are equal, the numerical value of skewness is zero. The value of this measure can vary between $+3$ and -3. Values larger than +1 or smaller than −1 are not frequent. When the value of the mode is not well defined, the following alternative formula can be used:

$$\text{sk.} = \frac{3(\text{mean} - \text{median})}{\text{S.D.}}.$$

The quartile measure of skewness. The formula is as follows:

$$\text{sk.} = \frac{Q_1 + Q_3 - 2\ \text{median}}{\frac{1}{2}(Q_3 - Q_1)}.$$

This measure can vary in value between +2 and −2.

The third moment measure of skewness. The formula is as follows:

$$\gamma_1 = \frac{\Sigma fd^3}{n\sigma^3}, \text{ where } d = \text{deviation from the arithmetical mean.}$$

(γ_1 is pronounced "gamma sub 1".)

However, a rather more complicated formula is used and deviations are made from an assumed average. This makes the calculations simpler. In the case of a grouped distribution, the deviations from the assumed average are in working units, but the value of the working unit is ignored in the calculation of this measure of skewness, since it occurs in both denominator and numerator.

There is no upper limit to this measure, but values greater than +2 or −2 are not frequent. These values would indicate a very great degree of skewness.

x	f	ξ	$f\xi$	$f\xi^2$	$f\xi^3$
5-10	1	−1	−1	1	−1
10-15	4	0	0	0	0
15-20	3	+1	3	3	3
20-25	1	+2	2	4	8
	9		4	8	10

Fig. 74.—Method of computation of the third moment measurement of skewness.

$$n = 9; \quad \Sigma f\xi = 4; \quad \Sigma f\xi^2 = 8; \quad \Sigma f\xi^3 = 10.$$

$$\gamma_1 = \frac{\dfrac{\Sigma f\xi^3}{n} - \dfrac{3\Sigma f\xi}{n} \times \dfrac{\Sigma f\xi^2}{n} + 2\left(\dfrac{\Sigma f\xi}{n}\right)^3}{\sigma^3}.$$

$$\gamma_1 = \frac{\dfrac{10}{9} - 3 \times \dfrac{4}{9} \times \dfrac{8}{9} + 2\left(\dfrac{4}{9}\right)^3}{\left(\sqrt{\dfrac{8}{9} - \left(\dfrac{4}{9}\right)^2}\right)^3} = 0{\cdot}2.$$

The distribution is positively skewed.

Questions

1. Write notes on two of the following:

(*a*) random sampling; (*b*) skewness; (*c*) business charts.

2. Compile a cumulative frequency table from the following data:

Weight (lb)	*No. of persons (hundreds)*
Under 100	1
100–109	14
110–119	66
120–129	122
130–139	145
140–149	121
150–159	65
160–169	31
170–179	12
180–189	5
190–199	2
200 and over	2

Draw a cumulative frequency curve, taking the upper limit at under 210. Estimate the median weight from the curve. *Institute of Statisticians.*

3. Explain the meaning of negative and positive skewness. Calculate the quartile coefficient of skewness of the frequency distribution in Question 2. *Institute of Statisticians.*

4.

Football Pools

Amount of stake	*Number of individuals*
6p and under 12p	35
12p " 24p	105
24p " 36p	76
36p " 48p	52
48p " 60p	35
60p " 72p	26
72p and over	12
	340

Calculate the standard deviation and the arithmetic mean of the given distribution. Assuming the mode is about 18p, estimate the skewness of the distribution.

5. Construct a histogram showing the distribution of football stakes given in Question 4. In your diagram, insert the median and the upper and lower quartiles. Find the ratio of the interquartile range to the total range. Take the last group to be 72p and under 84p.

6. What is meant by skewness and how is it measured? From the frequency distribution table given below, calculate the third moment measure of skewness.

X	*Frequency*
5–10	9
10–15	14
15–20	21
20–25	17
25–30	8
30–35	2
35–40	1
	72

7. (*a*) Explain and illustrate graphically the relationship between the mean, median and mode in (*i*) a positively skewed frequency distribution and (*ii*) a negatively skewed distribution.

(*b*) Calculate the coefficient of skewness for a frequency distribution with the following values: mean = 10; median = 11; standard deviation = 5. What does your answer tell you about this frequency distribution?

Association of Certified Accountants.

CHAPTER 21

CORRELATION

The meaning of correlation. If two quantities vary in such a way that movements in one are accompanied by movements in the other, these quantities are correlated. Many examples will come to mind. An increase in the amount of rain will be accompanied by an increase in the sales of umbrellas; older men usually have older wives than younger men; an increase in rainfall will up to a point be accompanied by an increase in the output of wheat per acre; and an increase in the issue of television licences is accompanied by a decrease in the number of cinema admissions. In each of the examples given, movements in one variable are accompanied by movements in the other. In the first three cases the movements were in the same direction, an increase of one variable was associated with an increase in the other variable. The correlation was positive. In the last case the movement of one variable was associated with a movement in the other variable in the opposite direction, an increase of one being accompanied by decrease in the other. The correlation was negative.

It is important to realise that although two quantities move in sympathy, changes in one do not necessarily cause changes in the other. Correlation is NOT synonymous with causation. The changes may both be due to a common cause. Thus, there is a considerable degree of correlation between deaths due to cerebral thrombosis and the number of old-age pensions. It cannot be said that old-age pensions cause cerebral thrombosis. The increase in both is due to a common cause, namely that people are living longer. There may be no connection whatsoever between the two quantities, apart from the fact that they just happen to move in sympathy. An example of such "nonsense correlations," as they are termed, is the considerable degree of correlation stated to exist between the number of births in the U.K. and the stork population of Scandinavia. Whether the movements indicate causation or not must be decided on other evidence than the degree of correlation.

The coefficient of correlation. This is the measurement of the degree of correlation between the variables. It will vary between +1 and −1. If there is perfect correlation, the coefficient will be +1 in the case of positive correlation (movements in the same direction) and −1 if there is negative, or inverse correlation. In the case of perfect correlation, if one quantity is known, the other can be calculated with certainty. Thus with a given voltage it is always possible to calculate the amount of electrical current when the resistance is known. There

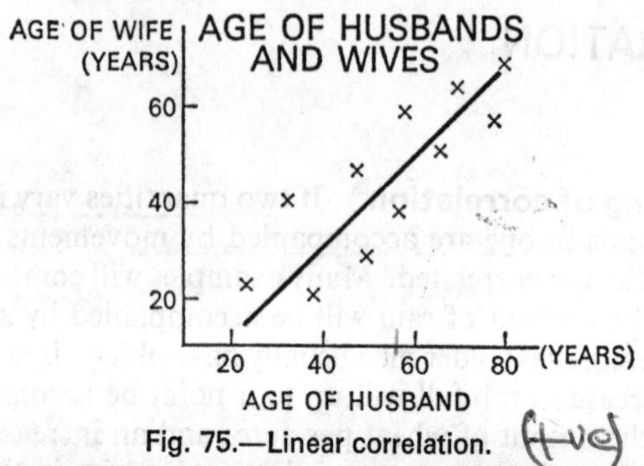

Fig. 75.—Linear correlation. (+ve)

would be no need to use statistical methods. In business, economic and sociological inquiries the coefficient of correlation is most likely to be less than 1. Given the value of one variable, the probable value and not the exact value of the other variable can be found.

Scatter diagrams. A useful method of investigating if there is any correlation between two variables is to draw a scatter diagram. In respect of each observation, the value of one variable will be measured along the *y*-axis, and the corresponding value of the other variable will be plotted along the *x*-axis.

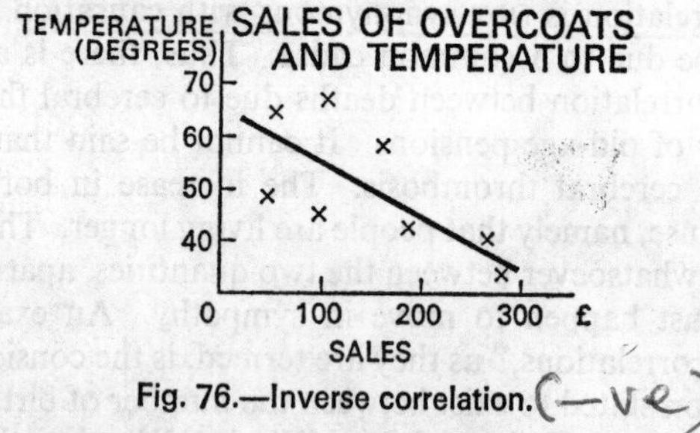

Fig. 76.—Inverse correlation. (−ve)

Examples of scatter diagrams. Fig. 75 shows in respect of each married couple the age of the husband plotted against the age of the wife. It will be seen that in most cases, but not in all, the older the husband, the older the wife. It will also be noted that for two husbands of the same age, the two wives may have different ages. However, there is a distinct tendency for the ages of wives and husbands to move in the same direction. The points plotted cluster along a straight line, a kind of average line. Since the points are near this line, there is a considerable degree of correlation. Since the line of "best fit" is a straight line, the correlation is linear. Further, since the movements of the variables are in the same direction, the correlation is positive. When the points are near the line, it is possible to draw it by inspection. A

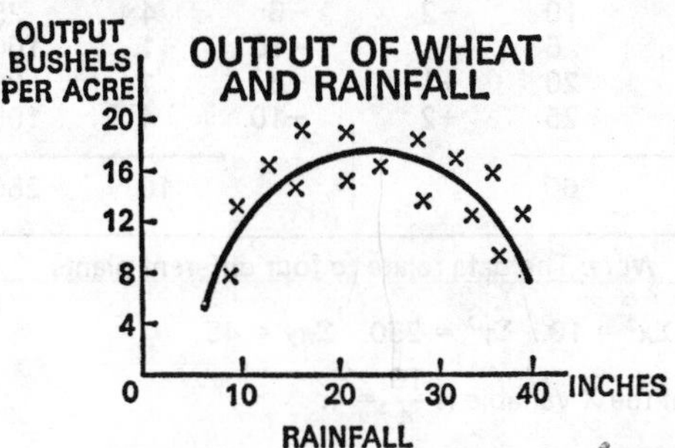

Fig. 77.—Curvilinear correlation.

good method is to stretch a piece of thread and try out the best position for the line. When, however, the points are too scattered to judge, recourse must be had to mathematical methods. Fig. 76 shows inverse linear correlation. Fig. 77 shows an example of a scatter diagram where a curve gives the best indication of the relationship between the

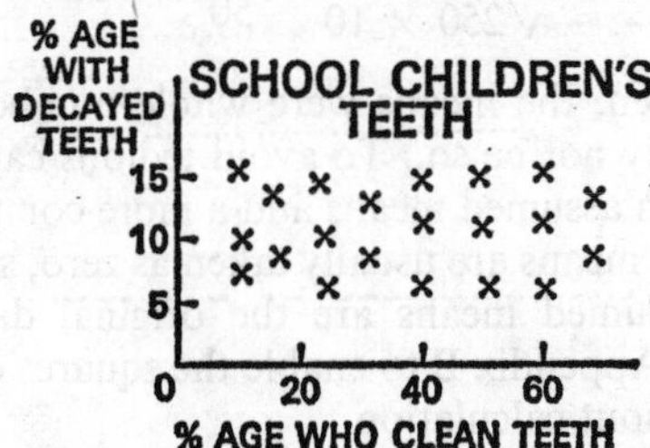

Fig. 78.—Zero correlation.

two variables. This is an example of curvilinear correlation. This kind of correlation is outside the scope of this book. Finally, Fig. 78 gives an example where there is no apparent correlation. The points do not appear to move in any direction.

Computation of the coefficient of correlation. The following formula only applies to linear correlation:

$$r = \frac{\Sigma xy}{\sqrt{(\Sigma x^2 \Sigma y^2)}}$$

where r is the coefficient of correlation, x is the deviation from the mean of one variable and y is the deviation from the mean of the corresponding variable.

Example 1. Find the coefficient of correlation between the number of weeks growth and the height in inches of a certain plant.

Plant	X (weeks)	Y (inches)	x	y	x^2	y^2	xy
1	2	10	−2	−5	4	25	10
2	3	5	−1	−10	1	100	10
3	5	20	+1	+5	1	25	5
4	6	25	+2	+10	4	100	20
	16	60			10	250	45

Note: The data relate to four different plants.

$\Sigma x^2 = 10. \quad \Sigma y^2 = 250. \quad \Sigma xy = 45.$

The mean of the X variable is $\frac{16}{4} = 4.$

The mean of the Y variable is $\frac{60}{4} = 15.$

The x's and y's are deviations from these means.
The column headed xy lists the products of these deviations.

Substituting these values in the above formula:

$$r = \frac{45}{\sqrt{250 \times 10}} = \frac{45}{50} = 0{\cdot}9.$$

In the example given, the means were whole numbers. In practice, they would probably not be so. To avoid tedious calculations deviations are taken from assumed means and a more complicated formula used. The assumed means are usually taken as zero, so that the deviations from the assumed means are the original data. A table of squares is given in Appendix B to enable the squares of the deviations to be obtained without calculation.

The necessary formula when deviations from assumed averages are taken is:

$$r = \frac{\frac{\Sigma XY}{N} - \frac{\Sigma X}{N} \times \frac{\Sigma Y}{N}}{\sqrt{\left\{\frac{\Sigma X^2}{N} - \left(\frac{\Sigma X}{N}\right)^2\right\}\left\{\frac{\Sigma Y^2}{N} - \left(\frac{\Sigma Y}{N}\right)^2\right\}}}.$$

This formula is not so formidable as it seems. The denominator is merely the product of the standard deviations of the two variables.

Using the data of the previous example, the coefficient of correlation will now be re-computed, using the formula, where the deviations are taken from assumed means. The assumed means are taken as zero for both variables.

Example 2.

X	Y	X^2	Y^2	XY
2	10	4	100	20
3	5	9	25	15
5	20	25	400	100
6	25	36	625	150
16	60	74	1,150	285

$\Sigma X = 16$; $\Sigma Y = 60$; $\Sigma X^2 = 74$; $\Sigma Y^2 = 1{,}150$; $\Sigma XY = 285$; $N = 4$.

Substituting these values in the formula

$$r = \frac{\frac{285}{4} - \left(\frac{16}{4}\right) \times \left(\frac{60}{4}\right)}{\sqrt{\left\{\frac{74}{4} - \left(\frac{16}{4}\right)^2\right\}\left\{\frac{1150}{4} - \left(\frac{60}{4}\right)^2\right\}}} = 0{\cdot}9 \text{ (as before).}$$

As many of the following operations can be carried out as is useful on either or both of the variables, and the resulting adjusted variables will have the same coefficient of correlation as the original variables.

(*a*) Add or subtract any amount to each of the values of the X variable. The same amount must be added to, or subtracted from, each item.

(*b*) Add or subtract any amount to each of the values of the Y variable, not necessarily the same as for the X variable The same amount must be added to, or subtracted from, each item.

(*c*) Divide or multiply each value of the X variable by any amount. Each item must be multiplied or divided by the same amount.

(*d*) Divide or multiply each value of the Y variable by any amount, not necessarily by the same amount as for the X variable. Each item must be multiplied or divided by the same amount.

If the pairs of values are plotted on a scatter diagram, the position of the points indicates the amount of correlation. The effect of the above operations will not affect their position; the adding or subtracting alters the position of the axes, not the relative position of the points, the multiplying or dividing alters the scale without altering the relative position of the points. Hence the coefficient of correlation is unchanged.

Example 3.

X	Y	$X-2$	$\frac{Y}{5}$	$X-2$	$\frac{Y}{5}-4$
2	10	0	2	0	−2
3	5	1	1	1	−3
5	20	3	4	3	0
6	25	4	5	4	1

Starting with the same figures as before, 2 was subtracted from each of the values of the X variable, and the Y variable divided throughout by 5. Then 4 was subtracted from each of the values of the adjusted Y variable.

Example 4.

X (*adjusted*)	Y (*adjusted*)	X^2	Y^2	XY
0	−2	0	4	0
1	−3	1	9	−3
3	0	9	0	0
4	1	16	1	4
8	−4	26	14	1

$\Sigma X = 8;\ \Sigma Y = -4;\ \Sigma X = 26;\ \Sigma Y^2 = 14;\ \Sigma XY = 1.$

$$r = \frac{\frac{1}{4} - \left(\frac{8}{4}\right) \times \left(\frac{-4}{4}\right)}{\sqrt{\left\{\frac{26}{4} - \left(\frac{8}{4}\right)^2\right\}\left\{\frac{14}{4} - \left(\frac{-4}{4}\right)^2\right\}}} = 0{\cdot}9 \text{ (as before).}$$

Test of significance of the coefficient of correlation. When the number of observations is small, it is not possible to place any reliance upon the value of "*r*."

If the value of "*r*" exceeds $\frac{3}{\sqrt{N-1}}$, where N is the number of observations, it can be assumed that the value of "*r*" is significant, that it does measure the amount of correlation.

Rank correlation. Instead of finding the coefficient of correlation between the actual values of two variables it is often sufficiently accurate to rank the variables in order of size; for example, instead of giving actual marks of (say) 6 pupils, give their positions, that is ranks, thus:

Pupil	A	B	C	D	E	F	
French	4	1	2	5	6	3	
Mathematics	1	3	2	4	5	6	
d (difference in ranks)	3	−2	0	1	1	−3	
d^2	9	4	0	1	1	9	$\Sigma d^2 = 24$

$$R = 1 - \frac{6\Sigma d^2}{n(n^2 - 1)} = 1 - \frac{6 \times 24}{6 \times 35} = 0{\cdot}3.$$

The coefficient of correlation for grouped data. From the figures below, calculate a coefficient of correlation between the age of the workers and the number of days lost through illness in a given year.

Illness of Employees in P.Q. Co.

Days lost \ Age	Under 30	30-39	40-49	50-59	60 and over	*f*	*y*	*fy*	*fy*²
1 or 2		2 **1*** 2		−2 **1** −2		**2**	**−2**	**−4**	**8**
3 or 4	2 **2** 4	1 **2** 2	**1**	−1 **1** −1		**6**	**−1**	**−6**	**6**
5 or 6	**1**	**3**	**4**			**8**	**0**	**0**	**0**
7 or 8		−1 **1** −1	**2**	1 **1** 1		**4**	**+1**	**4**	**4**
9 or 10			**3**			**3**	**+2**	**6**	**12**
over 10		−3 **1** −3			6 **1** 6	**2**	**+3**	**6**	**18**
f	**3**	**8**	**10**	**3**	**1**	**25**		**6**	**48**
x	−2	−1	0	+1	+2			Σ*fy*	Σ*fy*²
fx	−6	−8		3	2	−9	Σ*fx*		
*fx*²	12	8		3	4	27	Σ*fx*²		
fxy	4	0		−2	6	8	Σ*fxy*		

Fig. 79.—Bivariate frequency table.

$$r = \frac{\dfrac{\Sigma fxy}{n} - \left(\dfrac{\Sigma fx}{n}\right)\left(\dfrac{\Sigma fy}{n}\right)}{\sigma_x \sigma_y} = \frac{\dfrac{8}{25} - \left(\dfrac{-9}{25}\right)\left(\dfrac{6}{25}\right)}{\sqrt{\left(\dfrac{27}{25} - \left(\dfrac{9}{25}\right)^2\right)\left(\dfrac{48}{25} - \left(\dfrac{6}{25}\right)^2\right)}} = 0{\cdot}3.$$

* The bold figures are frequencies. The figures on the top left of these frequencies are the product of x and y (the deviations from assumed means in working units). The figures on bottom right are the product of the frequency and xy.

Questions

1. The following expenditure per child on clothing was obtained from an investigation carried out in York in 1950. Calculate the mean expenditure and the standard deviation.

Values to the nearest £

35 28 17 16 33 14 22 35 18 15 18 19 25 25 17 24
18 16 22 29 30 31 19 26 14 54 31 17 51

Source: *Poverty and the Welfare State.*
London Chamber of Commerce and Industry.

2. Draw a scatter diagram using the data given in Question 1 and the following expenditure on mothers' clothing. Hence, or otherwise, discuss the correlation between the two series of expenditure.

Expenditure on mothers' clothing
Values to the nearest £

46 25 23 47 32 30 11 35 15 28 14 22 51 28 33 43
26 16 31 30 28 35 22 29 45 27 19 23 41

Source: *Poverty and the Welfare State.*
London Chamber of Commerce and Industry.

3.

Merchandise: average value 1948 = 100
Average value

	1947	1948	1949	1950	1951	1952	1953
Imports	90	100	102	115	150	149	132
Exports	92	100	103	108	125	134	130

Discuss the amount of correlation between the average values of exports and imports by means of the scatter diagram and the correlation coefficient.

4. Show the following two series graphically so that any association between them is clearly brought out. If they are related, give an estimate of the relationship.

Series A	92	95	98	103	102	98	106	103	93	96
Series B	96	88	94	93	98	99	101	103	100	97

Series A	94	88	89	99	100	107	94	98	92	102
Series B	98	94	94	99	104	105	93	96	94	100

Chartered Institute of Transport.

5.

American Railroads

	Labour cost % of gross revenue	*Average journey per passenger (miles)*
1936	42·9	45·7
1937	44·8	49·6
1938	46·5	47·8
1939	44·1	50·3
1940	43·2	52·5
1941	41·1	60·5
1942	37·8	80·4
1943	36·9	99·6
1944	38·7	105·0
1945	41·4	102·9
1946	52·1	81·9
1947	47·6	65·3
1948	46·9	64·1
1949	48·9	63·3
1950	46·2	65·3
1951	48·2	71·5
1952	47·9	72·4
1953	47·5	69·3

How would you establish the existence of any link between the two series given above? How would you measure the degree of linkage and how can you test the significance of your results?

Institute of Company Accountants.

6.

	1966–7	1967–8	1968–9	1969–70
		(millions)		
Number of inland trunk calls	205	217	226	235
Number of inland telegrams	53	47	43	42

Draw the scatter diagram and calculate a coefficient of correlation.

Explain carefully, by means of the scatter diagram (or otherwise) the meaning of correlation. What does the correlation coefficient measure?

7. The following table shows the trend of cinema admissions and the growth of television licences in the Sutton Coldfield area during 1950–52.

		Cinema admissions (thousands)	*Television licences per 1,000 population*
1950	3rd quarter	10,025	24
	4th "	9,924	37
1951	1st "	9,814	52
	2nd "	9,726	64
	3rd "	9,632	69
	4th "	9,505	81
1952	1st "	9,405	98
	2nd "	9,271	101

Calculate the coefficient of correlation and comment on the result.

Institute of Statisticians.

8. Calculate a coefficient of correlation for the following two sets of figures. How would you interpret your result?

	Unemployment in Great Britain: monthly averages (000)	*Company's annual turnover* (£000)
1954	285	274
1955	232	297
1956	257	284
1957	312	317
1958	457	335
1959	475	386
1960	360	355

9. Two judges in a contest were asked to rank 8 candidates in their order of preference. The rankings were as follows:

Candidate	*Judge A*	*Judge B*
One	5	4
Two	2	5
Three	8	7
Four	1	3
Five	4	2
Six	6	8
Seven	3	1
Eight	7	6

Find the coefficient of rank correlation and say to what extent there is agreement between the two judges.

10. (*a*) Calculate the coefficient of correlation for the following data:

Firm	*Annual percentage increase in advertising expenditure*	*Annual percentage increase in sales revenue*
A	1	1
B	3	2
C	4	4
D	6	4
E	8	5
F	9	7
G	11	8
H	14	9

(*b*) What is the purpose of finding the correlation coefficient and what does its value indicate in respect of the above data on advertising expenditure and sales revenue?

Association of Certified Accountants.

COVARIANCE:

A very crude measurement of association ~~of association~~ is obtained by testing whether the sum of the positive products outweigh the sum of the negative products in the case under consideration. The sum of the products of the deviations is shown symbolically by $\Sigma(x-\bar{x})(y-\bar{y})$. This measure would not be adequate by itself as it would depend on the NUMBER of patterns as well as on the degree of ~~fee~~ association of the variables. To adjust for this we divide by the number of pairs of observations (n). The measure thus arrived at is called the CO-VARIANCE of x and y and is written

$$\text{Covariance} = \frac{\Sigma(x-\bar{x})(y-\bar{y})}{n}$$

CHAPTER 22

REGRESSION

The line of "best fit." In Chapter 21 the subject of linear correlation was discussed and Fig. 75 gives an example. It was there shown that the points giving the ages of husbands against the corresponding ages of their wives fell either side of a straight line and that this line was described as a line of "best fit"—a kind of average line.

The object of this chapter is to describe how this line can be drawn, to show how its formula can be obtained and to see, further, what is meant by this line of "best fit."

The equation $y = bx + a$. In Chapter 12, dealing with graphs, it was pointed out that x is the independent variable, that the value of y will depend upon the values given to x and that y is the dependent variable. Provided we know a and b, we can find the value of y for any given value of x. The following equations are examples of equations of the form $y = bx + a$.

$$y = 1{\cdot}2x - 2$$
$$y = -6x + 4$$
$$y = 0{\cdot}8x + 0{\cdot}7.$$

All these equations when shown in the form of graphs will be straight lines. The graph of the first equation would cut the y-axis at a point -2 units below the origin. This is fairly obvious, since when $x = 0$, $y = -2$. Similarly, the graph of the second equation cuts the y-axis at a point 4 units above the origin and in the third example 0·7 units above the origin. In general, the graph of the equation $y = bx + a$ will cut the y-axis a units above the origin (or, if a is a minus quantity, below).

In the first of the equations just given the value of y becomes 1·2 units more with every increase in the value of x by 1. 1·2 is the value of the slope of the line (also known as the tangent of the angle of slope). In the second equation the slope is negative, that is, as shown in Fig. 76, an example of inverse correlation (as x gets bigger y gets smaller). In general, the graph of the equation $y = bx + a$ has a slope whose value is b.

In Fig. 80 (*a*) the line shown has the equation $y = 1{\cdot}1x + 0{\cdot}7$.

It can be seen that the line cuts the y-axis at a point 0·7. The slope can be calculated as follows:

when $x = 3, y = 4$

when $x = 5, y = 6{\cdot}2$

Difference in value of $y = 2{\cdot}2$

Difference in value of $x = 2$

$$\text{Slope} = \frac{2{\cdot}2}{2} = 1{\cdot}1.$$ (b)

The regression equation. This is the equation which gives the relationship between variables when there is not a "unique" relationship between them. Thus, in the case of husbands' and wives' ages, given the age of the husband, it is not possible to say what the age of his wife will be, but it will be possible to estimate by giving the average age of wives whose husbands are of a given age. When the correlation is linear the regression equation will be of the form

$$y = bx + a.$$

The husband's age could be represented by x. It would be necessary to find the value of a and b; then for any given age of the husband the average age of wives of men of that age could be calculated. In any particular case the actual age of the wife might be different from the computed value. The sum of the deviations between actual and computed ages would be "nil." The line is therefore an "average" value line. This is what is meant by being a line of "best fit."

The regression of y on x. So far we have considered the case of "given x, what is the value of y?"; x is independent. This is also known as the regression of y on x, and b in the equation $y = bx + a$ is the regression coefficient. It shows that y changes b times as fast as x.

The regression of x on y. It is also possible to make y independent. Given the age of a wife, what is the age of her husband? The regression equation in this case is

$$x = b_1 y + a_1.$$

It is important to note that this line will *not* coincide with the regression equation of $y = bx + a$ unless there is complete correlation between the two variables, and in this case, of course, there would be no deviations from the regression line and the points of the scattergram would all lie on the straight line.

Calculations to establish regression equations.

Example 1.

Number of men employed on project	*Total output in units*
x	y
1	1
2	3
3	5
4	6
5	5

Iti s required to find:

(*a*) the regression of y on x;
(*b*) the regression equation of y on x;
(*c*) the regression of total output on number of men employed;
(*d*) the regression line $y = bx + a$.

These are four ways of saying the same thing. It is necessary to find b and a. Then given any number of men employed, the output can be estimated. It is also proposed to find the equation of $x = b_1y + a_1$ so that, given any output, we can estimate the number of men required on the project.

The calculations required are as follows:

x	y	x^2	y^2	xy
1	1	1	1	1
2	3	4	9	6
3	5	9	25	15
4	6	16	36	24
5	5	25	25	25
15	20	55	96	71

$\Sigma x = 15$; $\Sigma y = 20$; $\Sigma x^2 = 55$.
$\Sigma y^2 = 96$ $\Sigma xy = 71$.
$n = 5$ (there are 5 observations,).

These are precisely the same figures that must be computed to find the coefficient of correlation. If only the regression of y on x is required it is not necessary to calculate Σy^2.

The regression equation of y on x. The equation required is of the form

$$y = bx + a.$$

It is required to find the values of b and a. This necessitates solving the following simultaneous equations:

(*a*) $\Sigma y = b\Sigma x + na$
(*b*) $\Sigma xy = b\Sigma x^2 + a\Sigma x$.

In the example we are dealing with these are:

$$20 = 15b + 5a \qquad (1)$$

$$71 = 55b + 15a \qquad (2)$$

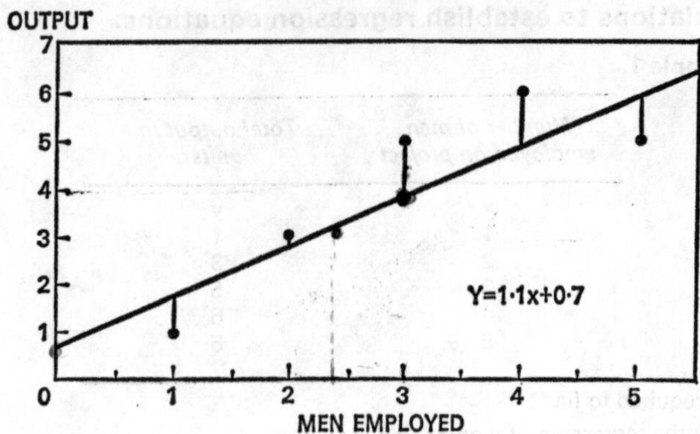

Fig. 80 (*a*).—Regression of *y* on *x*.

Multiplying (1) by 3,

$$60 = 45b + 15a \qquad (3)$$

Subtracting (3) from (2),

$$11 = 10b$$

$$\therefore\ b = 1{\cdot}1.$$

Substitute $b = 1{\cdot}1$ in equation (1)

$$20 = 16{\cdot}5 + 5a$$

$$\therefore\ a = \frac{3{\cdot}5}{5} = 0{\cdot}7.$$

The required equation is $y = 1{\cdot}1x + 0{\cdot}7$ (see Fig. 80 (*a*)).

The regression equation of *x* on *y*. The equation required is of the form

$$x = b_1 y + a_1.$$

It is required to find the values of b_1 and a_1. This necessitates solving the following simultaneous equations:

(*a*) $\Sigma x = b_1 \Sigma y + n a_1$
(*b*) $\Sigma xy = b_1 \Sigma y^2 + a_1 \Sigma y.$

In the example we are dealing with these are:

$$15 = 20b_1 + 5a_1 \qquad (1)$$

$$71 = 96b_1 + 20a_1 \qquad (2)$$

Multiplying (1) by 4,

$$60 = 80b_1 + 20a_1 \qquad (3)$$

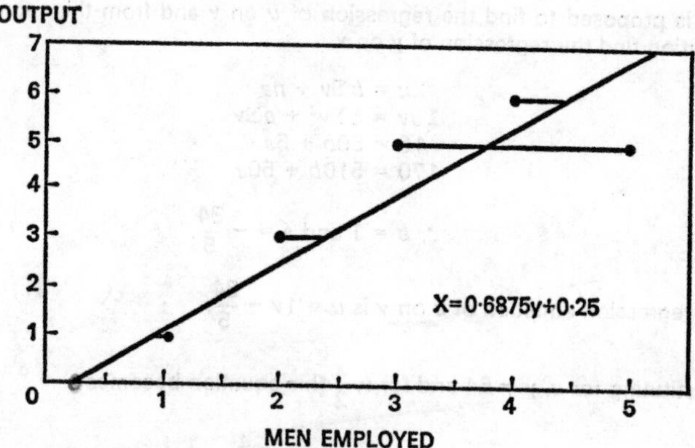

Fig. 80 (*b*).—Regression of *x* on *y*.

Deducting (3) from (2),

$$11 = 16b_1$$

$$\therefore\ b_1 = \frac{11}{16} = 0{\cdot}6875.$$

Substituting for b_1 in (1),

$$a_1 = \frac{15 - 20 \times 0{\cdot}6875}{5} = 0{\cdot}25.$$

The required regression equation is

$$x = 0{\cdot}6875y + 0{\cdot}25 \text{ (see Fig. 80 (}b\text{)).}$$

Changing the regression equation. Compare the variables *y* and *x* in the first two columns in Example 2 below with the variables *u* and *v* in the next two columns.

The variable *u* is obtained by subtracting 64 from the variable *y* and *v* is obtained by dividing *x* by 2. In Chapter 21 on correlation it was shown that the coefficient of correlation between *y* and *x* and between *u* and *v* were the same. The regression of *y* on *x*, however, is NOT the same as the regression of *u* on *v*. The origin and the rate of change have altered.

Example 2.

y	x	$u(y-64)$	$v\left(\frac{x}{2}\right)$	uv	v^2
68	18	4	9	36	81
64	16	0	8	0	64
67	20	3	10	30	100
69	24	5	12	60	144
68	22	4	11	44	121
		16	50	170	510

It is proposed to find the regression of u on v and from this regression equation find the regression of y on x.

$$\Sigma u = b\Sigma v + na \qquad (1)$$
$$\Sigma uv = b\Sigma v^2 + a\Sigma v \qquad (2)$$
$$16 = 50b + 5a \qquad (1)$$
$$170 = 510b + 50a \qquad (2)$$

$$\therefore\ b = 1 \text{ and } a = -\frac{34}{5}.$$

The regression equation of u on v is $u = 1v - \frac{34}{5}$.

Substituting for u, $y - 64$ and for v, $\frac{x}{2}$ this equation becomes

$$y - 64 = \tfrac{1}{2}x - \frac{34}{5}$$

$$\therefore\ y = \tfrac{1}{2}x + 57\tfrac{1}{2}.$$

This is the regression equation of y on x.

The regression coefficient. The slope of the regression line is known as the regression coefficient. It is the value of b and b_1 in the regression equations.

Regression and correlation. Figs. 80 (*a*) and 80 (*b*) show the graphs of the two regression lines. If they were superimposed it would be seen that the angle between them is very small, denoting a great deal of correlation between the two variables.

The coefficient of correlation is in fact the geometric mean of the two regression coefficients, and is therefore $\sqrt{1{\cdot}1 \times 0{\cdot}6875} = 0{\cdot}87$.

The following calculations will show that the deviations from the line of "best fit" add up to nil. Actual measurements on the graph will show the same thing.

Example 3.

$$y = 1{\cdot}1\,x + 0{\cdot}7$$

x	y (*actual*)	y (*computed from equation*)	*Deviation*
1	1	1·8	−0·8
2	3	2·9	+0·1
3	5	4·0	+1·0
4	6	5·1	+0·9
5	5	6·2	−1·2

$$x = 0{\cdot}6875y + 0{\cdot}25$$

y	x (actual)	x (computed from equation)	Deviation
1	1	0·9375	+0·0625
3	2	2·3125	−0·3125
5	3	3·6875	−0·6875
6	4	4·3750	−0·3750
5	5	3·6875	+1·3125

Alternative method of calculating the regression equation.

$$y = bx + a$$

$$b = \frac{\Sigma xy - \dfrac{\Sigma x \times \Sigma y}{n}}{\Sigma x^2 - \dfrac{(\Sigma x)^2}{n}}$$

$$= \frac{71 - \dfrac{15 \times 20}{5}}{55 - \dfrac{15^2}{5}}$$

$$= \frac{11}{10} = 1{\cdot}1 \text{ (as before).}$$

$$a = \frac{(\Sigma x)(\Sigma xy) - (\Sigma y)(\Sigma x^2)}{(\Sigma x)^2 - n(\Sigma x^2)}$$

$$= \frac{15 \times 71 - 20 \times 55}{15 \times 15 - 5 \times 55}$$

$$= \frac{-35}{-50} = 0{\cdot}7 \text{ (as before).}$$

Limits in the interpretation of a regression line. Interpolation between the extreme values of the observations given is normally quite safe. However, *extrapolation outside* the extreme values is often quite meaningless or dangerous. As an example of this, according to our equation, when no men are employed output is 0·7 units.

Questions

1. Plot the values in the following table on squared paper, and join them with a smooth curve:

x	y
1	10
2	19
3	27
5	40
6	45
7	49
8	52
9	54
11	55

(*a*) From your graph estimate: (i) the value of y when $x = 4$; (ii) the value of y when $x = 10$; (iii) the maximum value of y, and the value of x which corresponds to this value of y.

(*b*) Calculate the value of a and b and draw on your graph the "line of best fit" $y = a + bx$. Using this line, what values of y do you get for $x = 4$ and $x = 10$? *Institute of Cost and Management Accountants.*

(*Note:* The maximum value of y on the smooth curve is greater than 55. It reaches its highest value between 9 and 11 and has already started to fall before it becomes equal in value to 55 at the point when $x = 11$.)

2. Calculate the values of m and k for the equation $y = mx + k$ to show the regression of profit per unit of output on output.

Output (thousands)	*Profit per unit of output (£)*
5	1·7
7	2·4
9	2·8
11	3·4
13	3·7
15	4·4

Estimate from your equation the profit per unit of output when there is an output of 10,500.

3. The following figures refer to 10 specimens of a light metal alloy. The first row of figures gives the measure of hardness and the second row the corresponding tensile strength. You are required to find an equation from which an estimate of the tensile strength can be determined from a known hardness (assume linear relationship).

Hardness	22·9	17·8	20·8	21·3	20·7
Tensile strength	4·4	4·6	5·6	4·2	3·2
Hardness	20·9	17·5	13·6	23·3	18·1
Tensile strength	4·7	3·8	4·5	4·2	5·2

Institute of Statisticians.

4. From the data below fit a regression line of the form $y = ax + b$.

Year	Sales (£000)
1968	1
1969	3
1970	5
1971	6
1972	5

On the assumption that sales will continue to grow in the same way, estimate sales for 1974 (to nearest £ thousand).

5. The following table shows the amount of electricity generated by steam plant in Great Britain in the ten years, 1945 to 1954, and the amount of coal used in generation:

Year	Electricity generated (x) (thousand million kilo-watt hours)	Coal used (y) (million tons)
	(x)	(y)
1945	37	23
1946	41	26
1947	43	27
1948	47	29
1949	49	30
1950	55	33
1951	60	35
1952	62	36
1953	65	37
1954	72	40

Find the equations of regression for y on x and x on y.

Institute of Statisticians.

6. From the figures below obtain a formula giving pig-iron production in terms of coal consumption.

Year	Coal consumed in blast furnaces (in units of 100,000 tons)	Pig iron produced (in units of 100,000 tons)
1956	14·51	7·59
1957	11·69	6·19
1958	7·11	3·77
1959	6·53	3·57
1960	7·37	4·14
1961	10·47	5·97
1962	10·79	6·42

7. Overall death rates and percentage of deaths from certain diseases in eleven countries.

Country	*Total number of deaths per 1,000 population*	*Percentage of deaths from cancer, and diseases of the heart and circulatory system*
Australia	9·7	48
Canada	9·3	44
Chile	17·2	16
Denmark	8·6	46
Germany	10·2	35
Italy	12·0	25
Israel	6·9	31
Japan	11·9	11
Netherlands	8·1	39
Portugal	12·8	17
Spain	12·9	25
Sums	119·6	337
Sums of squares	1,383·5	11,979
Sums of cross-products	3,403·4	

Source: The Determinants and Consequences of Population Trends, U.N. 1954.

Either: (i) Explain what is meant by a regression line. Explain also what is measured by a regression coefficient.

For the above data obtain a linear equation from which the percentage of deaths from the specified causes could be predicted given the total number of deaths per 1,000 of the population; interpret the result.

Or: (ii) Calculate from the above data the coefficient of correlation and interpret your result. Explain what purpose is served by (*a*) the correlation coefficient, and (*b*) the regression coefficient.

B.Sc. (Sociology)—*London University.*

CHAPTER 23

PROBABILITY AND SIGNIFICANCE

Statistical probability. If an action can have any one of n equally likely results, and of these r would produce an event E, the probability of E is defined as $\frac{r}{n}$.

Example 1. From a bowl containing 100 red balls and 50 white balls, a ball is drawn at random. What is the probability it will be red?

If the ball is drawn at "random" (remember the technical sense of the word random) the action can have 150 equally likely results—any one of the 150 balls can be chosen. One hundred of these results would produce a red ball. The probability of drawing a red ball is therefore $\frac{100}{150} = \frac{2}{3}$.

Measurement of probability. It will be seen from the definition that probability can vary between 1 (that is certainty) and 0 (certainty that the event will *not* happen). Sometimes this probability is expressed as a percentage. The probability of obtaining a red ball in the example just given is 0·67 or 67 per cent. In betting language, the odds are two to one that a red ball will turn up.

Multiplication rule of probability. If there are a number of events and they are *independent* the probability of their *all* occurring is the product of their separate probabilities.

Example 2. A die is thrown into the air and a card is drawn from a well-shuffled pack of cards. What is the probability of getting *both* an ace *and* a six uppermost on the die?

Both these events are independent (the throwing of the die cannot affect the drawing of the card). The probability of *both* events occurring is therefore the product of their probabilities.

Probability of drawing an ace is $\frac{4}{52} = \frac{1}{13}$

(of the 52 cards 4 are aces).

Probability of a six uppermost $= \frac{1}{6}$

(of the six sides one is a six).

Probability of both events $= \frac{1}{13} \times \frac{1}{6} = \frac{1}{78}$.

Multiplication rule applied to conditional probability. When events are *dependent*, then the probability of any one event occurring depends upon whether the others have occurred.

Example 3. From a bowl containing 100 red balls and 50 white balls two balls are drawn at random. What is the probability they will both be red?

The probability of the first ball being red is $\frac{100}{150}$ (see Example 1). When the first ball is drawn and *if it is red*, there will be 99 red balls and 50 white balls left in the bowl. The probability of the second ball being red is therefore $\frac{99}{149}$. The probability of the first ball being red *and* the second ball being red is therefore

$$\frac{100}{150} \times \frac{99}{149} = \frac{198}{447}.$$

Addition rule of probability. If there are a number of events and they are *mutually exclusive* (that is, if any one happens the others cannot), then the probability of any one of them happening is the sum of their probabilities.

Example 4. A six-sided die is thrown into the air. What is the probability that *either* a three *or* a five will fall uppermost?

Both these events are *mutually exclusive.* (If a three turns up a five cannot, and if a five turns up a three cannot.) The probability of *either* a three *or* a five turning up is therefore the sum of their separate probabilities.

Probability of a three $= \frac{1}{6}$.

Probability of a five $= \frac{1}{6}$.

Probability of *either* three *or* five $= \frac{1}{6} + \frac{1}{6} = \frac{1}{3}$.

The binomial distribution. Let us suppose there is a large box containing a very large number of counters of various colours and that one-half of them are red. A random sample of three is taken, and it is required to know the probability of getting no red counters, one red counter only, two red counters only and three red counters. There are no other possibilities, and the sum of these probabilities must therefore be one.

First, and this is very important, when the first counter is taken from the box the probability of its being red is $\frac{1}{2}$, but the probability of the second being red is *not* $\frac{1}{2}$ (the proportion of reds in the box will have changed). However, if the number of counters in the box is very large we can ignore the fact that the probability has changed and treat the probability as constant.

The following are all the ways in which the sample of three counters can be drawn:

Example 5.

Combination	*1st Counter*	*2nd Counter*	*3rd Counter*
1	Red	Red	Red
2	Red	Red	Other
3	Red	Other	Red
4	Red	Other	Other
5	Other	Other	Other
6	Other	Red	Other
7	Other	Other	Red
8	Other	Red	Red

Each of these combinations is equally likely to be drawn and they all have the same probability, namely $\frac{1}{8}\left(\textit{i.e.}\ \frac{1}{2} \times \frac{1}{2} \times \frac{1}{2}\right)$.

There are eight ways in which we can draw these counters, and they are all equally likely. Of these, one way will give no red counters (combination 5); the chance or probability of there being no red counters is therefore $\frac{1}{8}$. Similar reasoning will show that the probability of getting two, and only two, red counters is $\frac{3}{8}$ (combinations 2, 3, 8).

In the form of a frequency table, these results can be shown as follows:

Example 6.

Number of red counters	*Frequency*	*Probability*
0	1	$\frac{1}{8}$
1	3	$\frac{3}{8}$
2	3	$\frac{3}{8}$
3	1	$\frac{1}{8}$
	8	1

Thus, of every eight random samples of three counters, one sample will have, on the average, no red counters; three samples will have one red counter, the other two not being red; three samples will have two red counters, the other not being red; and one sample will have three red counters.

This same result can be obtained by expanding the binomial $(q + p)^n$, where n is the number in the sample, p the probability of a "success" (in the example just given, the obtaining of a red counter), q the probability of a "failure" (in the example just given, the probability of getting a counter of a colour other than red).

It follows that $q + p = 1$. In the example just given, the probability of getting a red, each time a counter is drawn, is $\frac{1}{2}$; the probability of not getting it is therefore $\frac{1}{2}$; the probability of getting a red plus the probability of not getting it must be one, that is complete certainty.

Expanding $(q + p)^n$ where $p = \frac{1}{2}$, $q = \frac{1}{2}$, $n = 3$, we have $\left(\frac{1}{2} + \frac{1}{2}\right)^3$ $= \frac{1}{8} + \frac{3}{8} + \frac{3}{8} + \frac{1}{8}$ as before.

This distribution is known as the *binomial distribution.*

Another example of the binomial distribution. Suppose there is a box of component parts, one-third of which are defective. What is the probability of two, and only two, defectives if a random sample of three is drawn?

For each component part drawn the probability of its being a defective is $\frac{1}{3}$. The probability of its not being a defective is therefore $\frac{2}{3}$. $q = \frac{2}{3}$, $p = \frac{1}{3}$. Expanding the binomial $\left(\frac{2}{3} + \frac{1}{3}\right)^3$, we get

$$\frac{8}{27} + \frac{12}{27} + \frac{6}{27} + \frac{1}{27}.$$

The probability of getting two defectives only is $\frac{6}{27}$. The complete distribution is as follows:

Example 7.

Number of defectives	*Frequency*
0	8
1	12
2	6
3	1
	27

It will be seen that dividing the frequency of each value of the variable by the sum of the frequencies, we have probabilities. *Frequencies are probabilities.*

Standard deviation of the binomial distribution. Taking figures in Example 7, we get:

Number of defectives (x)	Frequency (f)	fx	fx2
0	8	0	0
1	12	12	12
2	6	12	24
3	1	3	9
	27	27	45

$$\text{S.D.} = \sqrt{\frac{45}{27} - \left(\frac{27}{27}\right)^2} = 0{\cdot}816$$

$$\bar{x} = \frac{27}{27} = 1.$$

This means that in random samples of 3 the number of defectives will vary, that the standard deviation will be 0·816 and that the average number of defectives will be 1.

These figures could have been obtained from the following formulae:

$$\text{S.D.} = \sqrt{pqn} \qquad \bar{x} = np.$$

In the above example $p = \frac{1}{3}$, $q = \frac{2}{3}$, $n = 3$.

$$\text{S.D.} = \sqrt{\frac{1}{3} \times \frac{2}{3} \times 3} = \sqrt{\frac{2}{3}} = 0{\cdot}816 \text{ (as before).}$$

$$\bar{x} = \frac{1}{3} \times 3 = 1 \text{ (as before).}$$

If it is desired to express the standard deviation of the proportion of the number in the sample (and the proportion in the sample is often used as an estimate of the proportion in the population), the formula to be used is

$$\text{S.D.} = \sqrt{\frac{pq}{n}}.$$

In the example just given, this would be $\sqrt{\dfrac{\frac{2}{3} \times \frac{1}{3}}{3}} = 0{\cdot}272.$

This equals the standard deviation of the number of defectives divided by the number in the sample.

Sample as an estimate of population. In the example just given, the proportion of defectives in the box was one-third; the average proportion in all the samples will also be one-third.

The proportion as given by the sample is taken as an estimate of the proportion in the population, *but* it is possible to measure the variation of this proportion from sample to sample. The difference between the proportion as given by the sample (the estimate of the population proportion) and the actual proportion in the population is the error of the proportion. The standard deviation of this error is known as the *standard error of the proportion,* and is in fact the standard deviation just given as the standard deviation of a proportion, namely $\sqrt{\frac{pq}{n}}$.

The "normal" distribution. As the number in the sample becomes larger, the binomial distribution becomes closer and closer to a "normal" distribution provided that pn is not less than 5; and if the sample is, say, 30 or more we shall be safe in treating it as if it were.

The "normal" distribution often occurs in biological data (*e.g.* heights of people), but as far as business statistics is concerned, its importance lies in the fact that when measurements and proportions vary by pure chance, and chance alone, the distribution of the measurements and proportions, provided the number of occurrences are sufficiently large, will be "normally" distributed.

This means that:

95 per cent of the items will vary in size between the mean size *plus* 1·96 standard deviations and the mean size *less* 1·96 standard deviations.

99·8 per cent will vary in size between the mean size *plus* 3·09 standard deviations and the mean size *less* 3·09 standard deviations.

It is possible to give probabilities for any number of standard deviations.

The frequency curve of a "normal" distribution is shaped as in Fig. 71.

Standardised deviates. Deviations from the mean in terms of standard deviation are known as standardised deviates. In Example 15 (*c*) the standardised deviate is 2·75. On referring to Appendix C—area under "normal" curve—it can be seen that the area from the mean to 2·75 standard deviations is 0·497. The area of the two tails is therefore 2(0·50 − 0·497) = 0·006. The probability of a difference of 2·75 standard errors or more is 0·6 per cent.

Significance. If we compare an actual observation with some theoretical expectation, the difference may be due to chance, for example, the fluctuations of random sampling (*i.e.* sampling error), or the difference may be *significant*, that is the theoretical expectation is wrong, *the null hypothesis* that there should be no difference is destroyed.

The problem is to determine whether the difference is significant. This can be done only in terms of probabilities.

Example 8. A coin is tossed 400 times and turns up heads 220 times. Is the coin biased?

Null hypothesis is that the coin is unbiased.

Standard error of number of heads on this hypothesis is

$$\sqrt{\frac{1}{2} \times \frac{1}{2} \times 400} = 10.$$

Mean number of heads on null hypothesis is 200 $\left(i.e.\ \frac{1}{2} \times 400\right)$.

Difference between actual observation and theoretical expectation

$$= 220 - 200 = 20.$$

$$\frac{\text{Difference}}{\text{Standard error}} = \frac{20}{10} = 2.$$

The probability of the number of heads turning up between 200 ± 20, *i.e.* between 180 and 220, is 95 per cent. The probability of its being as great or greater than 220 or as little as or less than 180 is 5 per cent.

The probability of getting a difference as great as or greater than 20, that is two standard errors, is 5 per cent. We can say that the difference is significant at the 5 per cent level of probability. This means that the difference is not due to chance, that the coin is biased, but that since we could get such a difference or greater, even though the coin were not biased in 5 cases out of 100, we should come to a wrong conclusion in 5 per cent of the cases.

At a probability level of 0·2 per cent we should say the coin was not biased. We should then require the difference to be as great as or greater than 30, *i.e.* 3 standard errors, before we considered the difference not due to chance.

If the difference had been 5 standard errors the result would have been highly significant—it would be practically certain that the coin was biased. The probability of getting such a large difference is very small. But improbable does not mean impossible.

Estimation of population proportion. The proportion in the population will not be known. The proportion as observed in the sample will therefore be used in the formula for the standard error. The estimation of the population proportion can only be given in terms of probabilities.

Example 9. In a random sample of 1,000 it was found that 300 people bought "X" type of detergent, 400 bought "Y" type and the remainder other types. What proportion of people bought detergents other than X or Y?

$$\text{Sample proportion} = \frac{1{,}000 - 300 - 400}{1{,}000} = 30\%.$$

$$\text{Standard error} = \sqrt{\frac{30 \times 70}{1{,}000}} = 1{\cdot}45\%.$$

$$\text{2 standard errors} = 2{\cdot}9\%.$$

95 per cent probability that the proportion of people who bought detergents other than X or Y = 30% ± 2·9%.

The maximum standard error of a proportion is when the proportion is $\frac{1}{2}$. If, therefore, the calculation is made as if this were the proportion there is a greater probability that the population proportion will be in the given range. This, in fact, would be necessary if the sample proportion were nil.

Fitting a normal distribution to sample data. In doing this we assume that the population is in fact normally distributed and that the mean and the standard deviation of the sample are the mean and standard deviation of the population (this entails a large sample).

Example 10. From a random sample of people it was found that their mean height was 66 inches and that the standard deviation was 2 inches. Give the distribution of heights of 100 people in class intervals of 1 inch.

Height (inches)	*Interval*	*Area proportion*	*Frequency*
60–61	$\bar{x} - 3\sigma$ to $\bar{x}$	0·50	1
61–62	$\bar{x} - 2\frac{1}{2}\sigma$ „ $\bar{x}$	0·49	1
62–63	$\bar{x} - 2\sigma$ „ $\bar{x}$	0·48	5
63–64	$\bar{x} - 1\frac{1}{2}\sigma$ „ $\bar{x}$	0·43	9
64–65	$\bar{x} - 1\sigma$ „ $\bar{x}$	0·34	15
65–66	$\bar{x} - \frac{1}{2}\sigma$ „ $\bar{x}$	0·19	19
66–67	$\bar{x}$ to $\bar{x} + \frac{1}{2}\sigma$	0·19	19
67–68	$\bar{x}$ „ $\bar{x} + 1\sigma$	0·34	15
68–69	$\bar{x}$ „ $\bar{x} + 1\frac{1}{2}\sigma$	0·43	9
69–70	$\bar{x}$ „ $\bar{x} + 2\sigma$	0·48	5
70–71	$\bar{x}$ „ $\bar{x} + 2\frac{1}{2}\sigma$	0·49	1
71–72	$\bar{x}$ „ $\bar{x} + 3\sigma$	0·50	1
			100

For the area proportions see Table in Appendix C.

Between the mean and the mean +2 standard deviations are 48% of the items or, as the table expresses it, 0·48 of the frequencies. In Example 10, that is from 66 to 70 inches. From 66 to 71 inches, that is from $\bar{x}$ to $\bar{x} + 2\frac{1}{2}$ standard deviations, the frequencies constitute 0·49 of the whole. Hence, the number of frequencies between 70 and 71 inches is 0·49–0·48 of the whole, that is 0·01 or 1%.

Significance of the mean (large samples). If a number of random samples are taken from a population, their means will vary from the mean of the population. These differences between the means of the samples and the population mean (sampling errors) will vary according to a normal distribution, provided that the samples are large enough, say, 30 or more. This is so, even though the population itself is (within limits) not normal.

The standard deviation of these errors, known as the *standard error of the mean*, is found by dividing the standard deviation of the population by the square root of the number in the sample.

Example 11. The average length of life of 400 lamps chosen at random was 900 hours and the standard deviation was 60 hours. The lamps were intended to have a life of 1,000 hours. Is the difference significant?

$$\text{Standard error of mean} = \frac{\text{S.D. of population}}{\text{Square root of number in sample}}$$

$$= \frac{60}{\sqrt{400}} = 3.$$

(The standard deviation of the sample taken as an estimate of the standard deviation of the population.)

$$\frac{\text{Difference in means}}{\text{Standard error}} = \frac{1{,}000 - 900}{3} = 33{\cdot}3.$$

The difference is highly significant. It is extremely unlikely that the lamps in the population have a life of 1,000 hours on the average.

There is a 99·8 per cent probability that the population mean is 900 ± 9 hours ($\bar{x} \pm 3$ standard errors).

Significance of the mean (small samples). When the sample is small, say less than 30, the method given above is inaccurate. We can no longer take the standard deviation of the sample as an estimate of the standard deviation of the population. Further, the sampling errors are not "normally" distributed. Special tables of the t distribution must be used and not the "normal" probability tables. (However, if the difference would not be significant using "normal" probability tables it will not be significant using t tables. The reverse does *not* hold good. A difference that is significant using normal probability tables will not necessarily be significant using t tables.)

Example 12. 5 boys had increased their weight after eating a certain product by the following amounts:

3lb; 4lb; 5lb; 6lb; 2lb.

Is the increase in weight significant?

$$\text{Mean increase in weight} = \frac{3 + 4 + 5 + 6 + 2}{5} = 4\text{lb}.$$

S.D. of sample $= \sqrt{\dfrac{9 + 16 + 25 + 36 + 4}{5} - 4^2}$

$= \sqrt{2}.$

Estimate of S.D. of population $= \sqrt{2} \times \sqrt{\dfrac{5}{4}} = \sqrt{2 \cdot 5} = 1 \cdot 6.$

(The estimate of the population standard deviation is obtained by multiplying the standard deviation of the sample by $\sqrt{\dfrac{n}{n-1}}$, known as Bessel's correction. Strictly speaking, this should also be applied to large samples, but the difference is negligible.)

$$\text{Standard error of mean} = \frac{1 \cdot 6}{\sqrt{5}} = 0 \cdot 71.$$

$$\frac{\text{Mean increase in weight}}{\text{Standard error of mean}} = \frac{4}{0 \cdot 71} = 5 \cdot 6.$$

Since this is a small sample, it will be necessary to consult tables of the *t* distribution.

The tables tell us (see Appendix D) that a difference as great as, or greater than, this has a probability of 0·5%. The increase is therefore significant and is evidence that the eating of this product caused an increase in weight.

Significance of the difference between means. It often happens that samples are from two different sources and it is required to know whether the difference in their means arises from there being a significant difference (that is, they are samples from different populations) or if the difference is a sampling difference (that is, the samples, although from different sources, really belong to the same population).

For large samples (say, over 30) the standard error of the difference of mean is:

$$\sqrt{\frac{\sigma_1^2}{n_1} + \frac{\sigma_2^2}{n_2}}$$

where n_1 is the number in the first sample, n_2 the number in the second; σ_1 is the standard deviation of the first sample and σ_2 the standard deviation of the second example.

The number of standard errors in the difference between the means determines whether the difference is significant in the same way as on page 153.

Confidence limits. Given the mean value in a sample (which must, of course, be a random sample), it is possible to give, with given probabilities of being right, limits within which the mean value of the population (from which the sample was taken) lies. These limits are known as *confidence limits.*

Example 13. A sample of 36 packets weigh on an average 8·4 oz each. The average weight of the whole of the packets from which the sample was taken can be given (*a*) *with given probabilities,* (*b*) *within given limits.*

DISTRIBUTION OF (LARGE) SAMPLE MEANS AND PROPORTIONS

(a) Reject null hypothesis at 5% level of significance

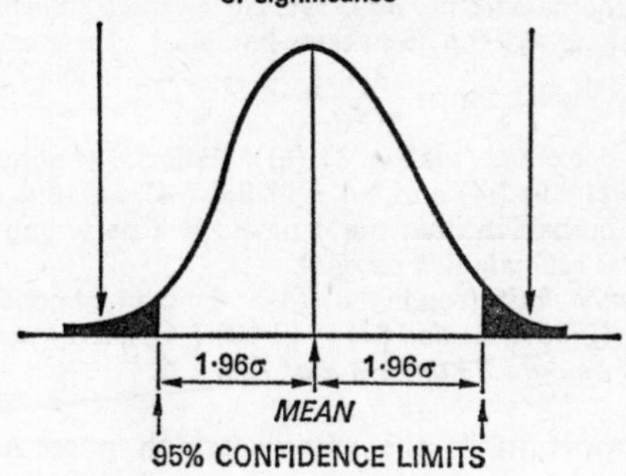

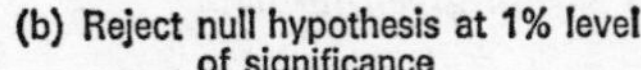

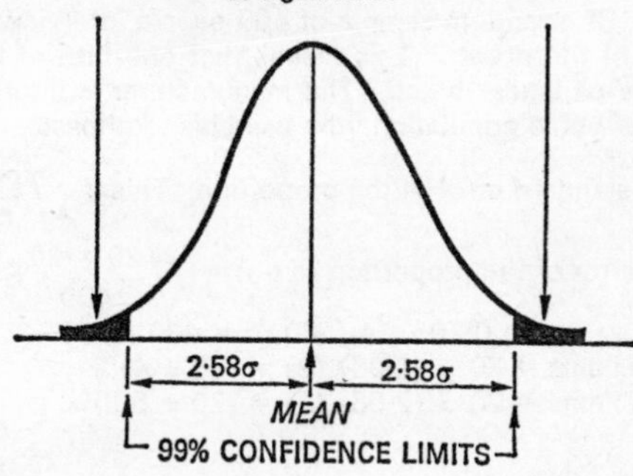

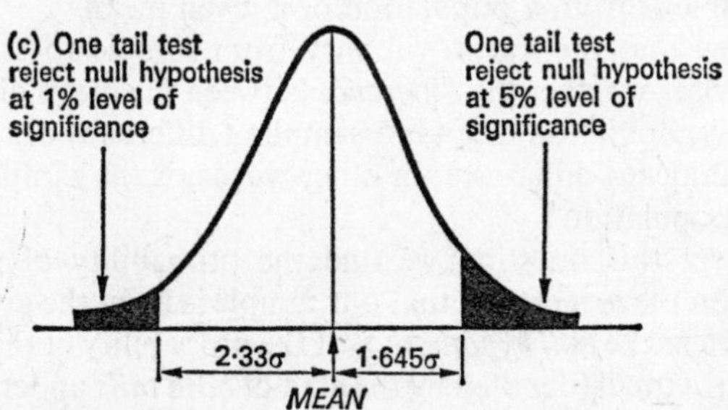

Fig. 81.—Confidence limits and tests of significance (normal distribution).

First, find the standard error of the mean. This is $\frac{\sigma}{\sqrt{n}}$. The standard deviation of the population is required. This is not known and so the standard deviation of the sample is taken as an estimate. This is sufficiently accurate where the sample is large, say, 30 or more. Let the standard deviation of the sample of the 36 packets be 2·4 oz. The standard error of the mean (s.e.$_{mean}$) will be then $\frac{2\cdot4}{\sqrt{36}} = \cdot4$.

1. *95% Confidence limits* (see Fig. 81 (a)). Estimate of population mean is between 8·4 + (1·96) (·4) and 8·4 − (1·96) (·4), *i.e.* (8·4 ± ·8) oz.

There is a 5% probability that the estimate will be wrong and a *95% probability that the estimate will be right.*

2. *99% Confidence limits* (see Fig. 81 (b)). Estimate of population mean is between 8·4 + (2·58) (·4) and 8·4 − (2·58) (·4), *i.e.* 8·4 ± (2·58) (·4). *This estimate will be right 99 times out of 100.*

Given the proportion in the sample which possesses a certain attribute, it is possible to estimate the proportion in the population from which the sample was taken that possesses this attribute (*a*) *with given probabilities*, (*b*) *within given limits.*

Example 14. Of a random sample of 400 people interviewed, 80 used a certain brand of toothpaste. This means that one-fifth of those in the sample used this particular brand. The manufacturer is interested in the proportion of the whole population who used his toothpaste.

First, find the standard error of the proportion. This is $\sqrt{\frac{pq}{n}}$.

The standard error of the proportion (s.e.$_p$) = $\sqrt{\frac{20 \times 80}{400}}\% = 2\%$.

(p = 0·20; q = 0·80; n = 400.)

95% Confidence limits = 20 ± (1·96) (2) = (20 ± 4)%.

99% Confidence limits = 20 ± (2·58) (2) = (20 ± 5·2)%.

Tests of significance (two tail test). Given the mean value in a sample, it is possible to say with given probabilities of being wrong if this sample is from a population of a given mean.

We know that the mean will vary from sample to sample. The question arises whether the *difference* between the mean of the sample and a given population mean is a sampling difference or a real difference, a significant difference; in other words, is the sample one from the given population?

To answer this question we find the probability of getting this *difference* on the *assumption* that our sample is from the given population (known as the *null hypothesis*). If the probability of this difference is sufficiently small (denoted by the area of *both tails* under the normal curve (see Figs. 81 (*a*) and 81 (*b*)), the argument is that the sample is unlikely to be from the given population, that is, our null hypothesis is rejected. The difference is significant, not a sampling difference.

Example 15. Suppose the 36 packets whose mean weight was 8·4 oz and whose standard deviation was 2·4 oz were taken from a population of packets whose mean weight was claimed by the packers to be (*a*) 9·0, (*b*) 9·25, (*c*) 9·5 oz. Would the packers' claim be justified?

First, find the standard error. This = ·4.

(*a*) Difference (often denoted by Z) $= \dfrac{9{\cdot}0 - 8{\cdot}4}{{\cdot}4} = 1{\cdot}5$ s.e.

Probability greater than 5% (see Fig. 81 (a)).

The null hypothesis that the sample was from the claimed population is accepted.

The difference is not significant at the 5% level. The packers' claim is probably justified.

(*b*) $Z = \dfrac{9{\cdot}25 - 8{\cdot}4}{{\cdot}4} = 2{\cdot}125$ s.e.

Probability less than 5%, but greater than 1% (see Figs. 81 (a) and 81 (b)).
Significant at 5% level but not at 1% level.
The packers' claim not justified at 5% level, but justified at 1% level.

(*c*) $Z = \dfrac{9{\cdot}5 - 8{\cdot}4}{{\cdot}4} = 2{\cdot}75$ s.e.

Probability less than 1% (see Fig. 81 (b)). (2·75 >2·58.)
Significant at 1% level.
Packers' claim not justified.

(There is less than one in a hundred chance, *i.e.* less than 1% chance, in being wrong and saying there is a significant difference when there is not since 2·75 >2·58.)

Given the proportion in the sample which possesses a certain attribute, it is possible to say, with given probabilities of being wrong, if the sample is from a population with a given proportion possessing this attribute.

Example 16. A manufacturer claimed that 25% of people used his toothpaste. In a sample of 400, 80 people used it. Was his claim justified?

Standard error of a proportion = 2%.

$Z = \dfrac{25 - 20}{2} = 2{\cdot}5$ s.e.

Significant at 5% level.
Claim probably not justified.

Tests of significance (one tail test). In the case of the two tail test we were concerned with the *probability of a difference* occurring. The probability was the area of the *two tails*. In the *one tail* test we are concerned with the *probability of the amount greater or the amount less occurring*.

Example 17. The sample of 36 packets had an average weight of 8·4 oz. The manufacturers claimed an average weight of (*a*) 9·0 oz, (*b*) 9·25 oz, (*c*) 9·5 oz. Is the average weight of the packets *less than* the manufacturers' claim?

(*a*) Underweight 1·5 s.e.

Probability greater than 5% (1·5 s.e. <1·645 s.e.).
Not significant at 5% level (see Fig. 81 (c)).
Average weight of packets probably not less than manufacturers' claim.

(*b*) Underweight 2·125 s.e.
Not significant at 1% level but significant at 5% level (see Fig. 81 (c)). (1·645 <2·125 <2·33.)
Average weight of packets probably less than manufacturers' claim.
(*c*) Underweight 2·75 s.e.
Probability less than 1%.
Average weight of packets, that is, in the population, less than weight claimed by manufacturers at 1% level.

The Poisson distribution. Given that a proportion (denoted by "p") of a population possesses some attribute and samples of size "n" are taken, then the number possessing this attribute will vary from sample to sample, but the average number in *all possible samples* will be "pn".

It has already been stated that where "pn" is not less than 5 and "n" is large, the binomial distribution approximates to the "normal" distribution Where, however, "p" is small and "n" is large and "*pn*" *is less than 5*, the binomial distribution approximates to the *Poisson* distribution.

Example 18. A manufacturer claims that 98% of equipment supplied is in conformity with the specification given. A random sample of 200 was taken to test the manufacturer's claim. Which distribution would the statistician use?

The proportion of equipment not up to standard was 0·02 (*i.e.* 2%); the sample size was 200; pn = 0·02 × 200 = 4. This means there will be on the average 4 pieces of unsatisfactory equipment in samples of 200. The appropriate distribution is the *Poisson*.

Properties of the Poisson distribution. The Poisson distribution occurs in many cases, *e.g.* the number of misprints per page in a book having a large number of pages, the number of telephone calls per minute at a switchboard The variable is always an integer—a whole number.

The mean value (denoted by "m") is the only information required to determine the distribution (The variance is also "m".)

The probability of obtaining x misprints per page, calls per minute or defectives in a sample when the Poisson distribution is appropriate is

$$\frac{m^x e^{-m}}{x!}$$

Values of e^{-m} are given for certain values of m in Appendix G. Values of x! (x factorial) are given for certain values of x in Appendix H.

Example 19. If the number of pieces of unsatisfactory equipment found in the sample of Example 18 were 6, test the manufacturer's claim at the 5% level.

$$P(6) = \frac{4^6 e^{-4}}{6!}$$
$$= \frac{4{,}096 \times 0{\cdot}01832}{720}$$
$$= 0{\cdot}10 \text{ (i.e. 10\%)}$$

As the probability of obtaining 6 defective pieces is greater than 5%, the manufacturer's claim is accepted.

Questions

1. The mean I.Q. of a sample of 1,600 children was 101. How likely is it that this was a random sample from a population with a mean I.Q. of 100 and a standard deviation of 16?

2. 40 cartons are taken at random from an automatic filling machine. Their mean net weight is 7·9 oz and the sum of the squares of the deviations from the mean is 0·308. Does the sample mean differ significantly from the intended weight of 8 oz?

3. The mean breaking strength of steel rods is given as 25,000 lb. The breaking strength of 120 rods were found to have a mean of 23,800 lb with a standard deviation of 560 lb. Is the complaint that the rods were not up to specification justified?

If only 60 rods had been tested and were found to have the same mean and standard deviation would the complaint still be justified?

4. A manufacturer makes a certain brand of soap and claims that 75 per cent of housewives prefer it to any other make. A random sample by 300 housewives contains 216 who do prefer the brand. Is the manufacturers' claim justified?

5. A machine making components to a nominal dimension of 3·000 inches is reset every morning. The first 36 components produced one morning have a mean of 2·988 inches with a standard deviation of 0·004 inches. Does this provide sufficient evidence that the machine is set too low?

6. A manufacturer claimed that at least 98 per cent of equipment supplied was in conformity with the specifications given. An examination of a sample of 240 pieces of equipment showed that 12 pieces were not in accordance with the specification. Test the manufacturers' claim at (*a*) 1 per cent, (*b*) 5 per cent levels of significance.

7. A random sample of 100 male students provided the following data:

Height in inches	*No. of students*
60–62	5
63–65	18
66–68	42
69–71	27
72–74	8
	100

From this data estimate (*a*) 95 per cent, (*b*) 99 per cent confidence limits of mean height of **all** male students.

8. In a random sample of 400 students, 25 per cent did not know what career they intended to adopt after they had graduated. What are (*a*) the 95 per cent, (*b*) the 99 per cent confidence limits of the percentages of students who do not know what their career will be?

9. The following data relate to measurements of components produced by two different machines. Is there any significant difference between the output of the two machines?

Machine	*Number measured*	*Average size*	*Standard deviation*
A	1,200	31 inches	5 inches
B	1,000	30 inches	4 inches

10. Suppose that 10 per cent of the units produced in a batch are defective; what is the likelihood that:

(*a*) one of them picked out at random passes the quality test;
(*b*) two out of a random sample of two are defective;
(*c*) this random sample of two contains only one defective?

Give reasons for your estimate and a general rule which could be applied in the same calculation for a sample of three.

Institute of Cost and Management Accountants.

11. A biscuit manufacturer produces half-pound packets of line X. A random sample of packets was weighed and the following figures recorded:

Net weight (oz)	*Number of packets*
less than 7·4	0
7·4 " 7·6	1
7·6 " 7·8	8
7·8 " 8·0	24
8·0 " 8·2	31
8·2 " 8·4	20
8·4 " 8·6	4
8.6 and more	2

(*a*) Calculate the mean and standard deviation of the weight per packet.
(*b*) With what degree of certainty can the manufacturer state "minimum net weight $7\frac{1}{2}$ oz"? *Institute of Cost and Management Accountants.*

12. The following figures relate to an experiment carried out in the brewing industry in order to test whether continuous brewing gives results comparable with those of batch brewing. Two samples were taken from each of the six brews; one was then processed by the continuous, the other by the batch method. Laboratory analysis gave the following results relating to sugar content:

Brew no.	*Processed*	
	Continuously	*By batch*
1	35	30
2	38	34
3	35	45
4	34	42
5	56	35
6	26	38

Show whether there is any significant difference between the results of the two methods of processing. Would the result of this calculation be different if you were told that the samples taken from Brew No. 5 were faulty and should be excluded from the experiment?

Institute of Cost and Management Accountants.

13. The following represents a random sample of daily sales (in £s) of a trader and refer to one line only:

120, 98, 114, 80, 131, 118, 130, 110, 105, 107.

Calculate his average daily sales of this line and the standard deviation.

If he could not replenish until the following morning, what value of stock should he hold to be 95 per cent sure that he would not run out that day? (Assume that the distribution is normal.)

Institute of Cost and Management Accountants.

14. A machine produces metal rods of nominal length 20 inches. The following results were obtained when the output was checked.

Length of rod (inches)		*Number of rods*
Not less than	*Less than*	
16·5	17·5	20
17·5	18·5	40
18·5	19·5	80
19·5	20·5	130
20·5	21·5	90
21·5	22·5	30
22·5	23·5	10

You are required to:

(*a*) calculate the average length of rod, and the standard deviation of these lengths;

(*b*) estimate within what limits you could reasonably expect the average length of rods in other batches to lie if production continues under the same conditions in batches of 400;

(*c*) plot the numbers given in the above table on squared paper, and join them with a smooth curve. Draw the ordinates at—

(i) mean + 1 standard deviation, and

(ii) mean + 2 standard deviations.

Shade in the area between these ordinates. If your curve were normal, what proportion of the whole area under the curve would the shaded area occupy?

(You may use the following information:

from $\bar{x} - \sigma$ to $\bar{x} + \sigma = 0{\cdot}683$
from $\bar{x} - 2\sigma$ to $\bar{x} + 2\sigma = 0{\cdot}955$.)

Institute of Cost and Management Accountants.

15. A machine produces metal parts and feeds them into a bin. An empty bin is placed by the machine after every 100 parts and the full bin wheeled away for inspection. After checking a large number of bins the inspector noted that per bin of 100 parts the average number of defective parts was 10 and the standard deviation of the number of defective parts was 3.

Assuming that production continues under the same conditions:

(*a*) if larger bins were used each holding 300 parts what would be—

(i) the average number of defective parts per bin,

(ii) the standard deviation of the number of defective parts per bin?

(*b*) how many parts must the bin be able to hold so that the standard deviation of the number of defective parts per bin is equal to 1 per cent of the total number of parts in the bin?

Assuming that owing to a change in the conditions of production the machine produces good and bad parts in random order in the proportion of 99 : 1 and that bins each holding 100 parts are used:

(*c*) estimate the proportion of bins which would contain no defective parts, and

(*d*) what would you expect the standard deviation of the number of defective parts to be?

Show how you arrive at your answers.

Institute of Cost and Management Accountants.

16. The manufacturers of brand "X" margarine held a tasting test to determine whether people could distinguish between their product and butter. Each subject was presented with 6 biscuits, 5 of which were spread with butter and one with brand "X" margarine, and asked to pick out the margarine.

Of 1,680 people tested, 345 picked out the margarine-spread biscuit correctly.

(*a*) What proportion of people would you expect to pick out the odd biscuit, if their choice were purely random?

(*b*) Could the results given above be purely due to chance or could some people tell brand "X" margarine from butter?

(*c*) How does the theory of the normal curve help in assessing the results?

Institute of Cost and Management Accountants.

17. (*a*) What are the more important characteristics of the Normal distribution?

(*b*) Discuss the importance of the mean and standard deviation in the use of the Normal distribution.

I.C.S.A. Part 1, Dec. 1975.

CHAPTER 24

CHI-SQUARED AS A TEST OF ASSOCIATION

Contingency tables. A contingency table consisting of m rows and n columns is an $m \times n$ contingency table. An example of a 3×2 table is shown in Fig. 82.

Political Affiliation of Employees

	Manual employment	Non-manual employment	Totals
Conservative	18	19	37
Labour	46	14	60
Liberal	11	13	24
Totals	75	46	121

(Hypothetical data)

Fig. 82.—A 3 × 2 table.

Each observation is classified in two ways; according to (*a*) political party, (*b*) type of employment. The above table shows, for example, that 46 out of a total of 121 observations consisted of people belonging to the Labour Party who were engaged in manual work.

The contingency table provides a means of testing if there is any association between the characteristics upon which the classification is based; in the above example, if there is any association between political party and type of work.

The "null" hypothesis. This is the hypothesis that there is no relationship between the characteristics. In the above example, the null hypothesis is that there is *no association* between political party and type of work. If this hypothesis is correct, the proportion of Conservatives who are engaged in manual work will be the same as the proportion of Conservatives among all the observations; that is, the expected number of Conservatives engaged in manual work $= \frac{37}{121}$ (the proportion of Conservatives in total) $\times$ 75 (the total engaged in manual employment) $=$ 22·9.

The other expected frequencies are obtained in a similar way. They are tabulated below:

	Manual employment	*Non-manual employment*	*Totals*
Conservative	22·9	14·1	37
Labour	37·2	22·8	60
Liberal	14·9	9·1	24
Totals	75·0	46·0	121

The totals must, of course, be the same as for the observed frequencies.

χ^2 (Chi-squared). This measures the difference between the expected frequencies and the observed frequencies, and is computed as follows:

$\chi^2 = \Sigma \dfrac{(O - E)^2}{E}$, where O is the observed frequency and E is the expected frequency in respect of each cell. In the above example

$$\chi^2 = \frac{(18 - 22{\cdot}9)^2}{22{\cdot}9} + \frac{(19 - 14{\cdot}1)^2}{14{\cdot}1} + \frac{(46 - 37{\cdot}2)^2}{37{\cdot}2} + \frac{(14 - 22{\cdot}8)^2}{22{\cdot}8} + \frac{(11 - 14{\cdot}9)^2}{14{\cdot}9} + \frac{(13 - 9{\cdot}1)^2}{9{\cdot}1} = 11{\cdot}1.$$

Degrees of freedom. The number of degrees of freedom for a contingency table $m \times n$ is $(m - 1)(n - 1)$. For the 3×2 table just given the number of degrees of freedom is

$$(3 - 1)(2 - 1) = 2 \times 1 = 2.$$

Degrees of freedom are denoted by υ (pronounced nu), n or by d.f. The table above has 2 d.f. This means that only 2 expected frequencies need be computed. The others are obtained by subtraction from marginal totals.

Yates's correction. In the case of a 2×2 table there is only one degree of freedom, and it is necessary to make an adjustment in the computation of χ^2. *This adjustment is made only in the case of one degree of freedom.* This is done as follows:

If $(O - E)$ is, say, $-3{\cdot}2$, then the adjusted $(O - E)$ is $-2{\cdot}7$ (0·5 is deducted from 3·2) and $(O - E)^2$ is $(-2{\cdot}7)^2$. If $(O - E)$ is, say, 3·2, then the adjusted $(O - E)$ is 2·7 (0·5 is again deducted from 3·2) and $(O - E)^2$ is $2{\cdot}7^2$. 0·5 is always taken from the absolute value of $(O - E)$ whether it is negative or positive.

The significance of chi-squared. If the null hypothesis is true, that is, there is no association between the two characteristics, yet there can still arise a difference between the expected frequencies and the observed frequencies. The greater χ^2 (which measures these differences), the smaller the probability of its occurring. The probability of χ^2 of any given amount also depends upon the number of degrees of freedom. In the case above, χ^2 was 11·1 with two degrees of freedom. Reference to tables of χ^2 (Appendix E) shows the probability of χ^2 being 11·1 or greater with two degrees of freedom is less than 0·005. This is such a small probability that the null hypothesis is rejected, and the evidence indicates there is a relationship between political party and type of employment.

Questions

1. *Relationship between aggressive mood and illness behaviour in a sample of male employees in a large company.*

Aggressive mood	*Frequency of illness behaviour*			*Total*
	Rare	*Intermediate*	*Frequent*	
Low	59	51	12	122
High	59	82	39	180
Total	118	133	51	302

Source. S. V. Kasl, *Journal of Chronic Diseases,* 1964.

On the assumption that the data are derived from a random sample, test the hypothesis that there is no association between aggressive mood and frequency of illness behaviour. Explain fully the meaning of your result.

B.Sc. (Sociology), University of London.

2. Explain briefly the purpose of the χ^2 (Chi-squared) test of association for contingency tables.

Change in Turnout of Electors and Swing to Labour in 95 Conservative-held non-rural Constituencies at the 1964 General Election

Change in turnout	*Swing*		*Total*
	4% or less	*more than 4%*	
2·1% fall or more	10	27	37
Less than 2·1% fall, or rise	29	29	58
Total	39	56	95

Source: H. B. Berrington, *Journal of the Royal Statistical Society, 1965.*

For the above data, test whether there is a significant association between change in turnout and swing to Labour. Interpret your result.

B.Sc. (Sociology), University of London.

3. The Acca Co. Ltd. tests all components built into its products to ensure that they have attained the standard of quality required. There are three checkers, A, B and C, who undertake this work. The table below shows the numbers of components accepted and rejected by the checkers when they tested three batches of component 1703 recently delivered to the factory. You are required to test the hypothesis that the proportions of components rejected by the three are equal and explain the significance of your conclusion.

Use the chi-squared (χ^2) distribution as a test of significance, at the ·05 level.

The Acca Co. Ltd.
Components quality test

	Checker A	*Checker B*	*Checker C*	*Total*
Accepted	44	56	50	150
Rejected	16	24	10	50

Association of Certified Accountants.

CHAPTER 25

ANALYSIS OF TIME SERIES

Time series. A time series has been defined as "data classified chronologically." It is a set of observations, for example, sales or production, over time (see Fig. 83).

Amigo Cement Co. Ltd. production (Thousand tons)

	1st Qr.	2nd Qr.	3rd Qr.	4th Qr.
1968	83	81	98	114
1969	124	113	115	152
1970	163	162	168	175
1971	191	180	184	197

Fig. 83.—A time series.

Characteristics of a time series. Two things are noticeable about the time series given in Fig. 83. First, there is a steady growth in production throughout the period. Secondly, there appear to be regular seasonal fluctuations, production declining during the second and third quarters of the firm's year.

A time series is the result of a number of movements. They are as follows:

1. A *basic trend.* This will be the long-term movement. In Fig. 83, it is a steady growth.
2. *Seasonal fluctuations.* These are deviations from the trend. They generally occur periodically, every quarter, week or month, according to the nature of the data.
3. *Catastrophic movements.* These are caused by unusual events, for example, floods, strikes, fires, etc.
4. *Cyclical fluctuations.* These are oscillatory movements superimposed on the trend. In the case of economic and industrial time series they will correspond to movements of the trade cycle.
5. *Residual variations.* These are the variations which are left

after having removed all other movements and fluctuations. They will include chance variations arising from a multiplicity of causes.

The analysis of time series consists in separating these constituent parts.

The uses of time series analysis. It enables forecasts to be made. If the trend line is continued into the future, as is shown by the broken line in Figs. 84, 85 and 86, a possible figure for some future

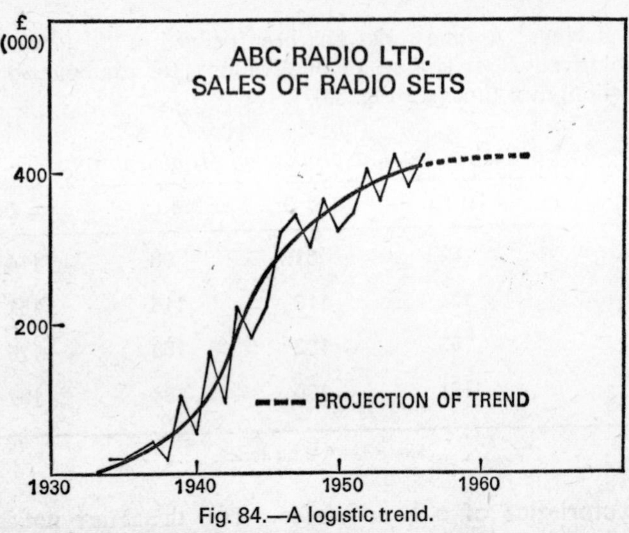

Fig. 84.—A logistic trend.

date is obtained. Statisticians are not fortune-tellers, and the figure obtained by extrapolation, *i.e.* by projecting the trend line, is made on the assumption that the trend will continue. Any unforeseen circumstances may invalidate the forecast. Other knowledge which may modify the projection must be taken into account. For the forecast to be of any value, the series from which it is obtained must be a long series. Despite the considerable number of snags, forecasts based on the analysis of time series are useful in planning the various budgets in budgetary control.

If seasonal fluctuations are known, it is possible to adjust holdings of stocks so that unnecessary stocks are not held, or, on the other hand, to avoid having insufficient. It is an essential aid to planning, permitting perhaps other commodities to be produced or handled so that fluctuations are "evened out."

Basic trends. Three of the more usual basic trends are shown in Figs. 84, 85 and 86. The logistic trend has been found to be appropriate in many cases of natural growth, in the growth of production of certain commodities and in many types of business activity. The trend consists of an increasing rate of growth, followed by a decreasing rate of growth, finally reaching a constant figure. In the case of sales, this would indicate that the market had reached saturation point.

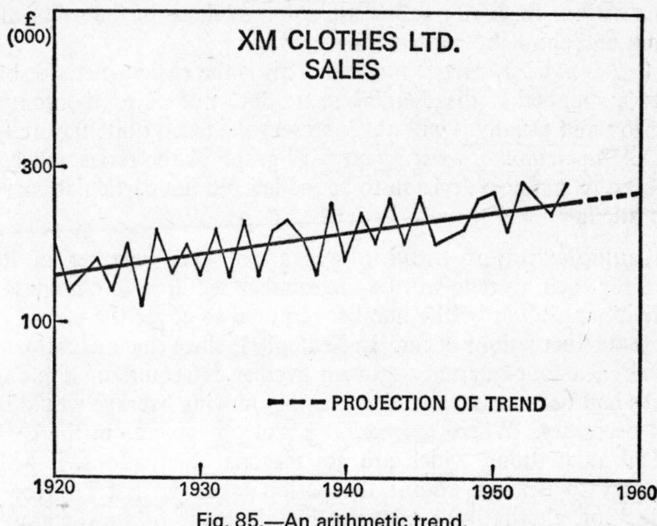

Fig. 85.—An arithmetic trend.

Fig. 85 shows an arithmetic trend. This means that the variable increases or decreases by a constant amount each year.

Fig. 86 shows a compound interest trend. Such a trend will appear on semi-logarithmic paper as a straight line. This means that the variable will increase or decrease by the same proportion each year.

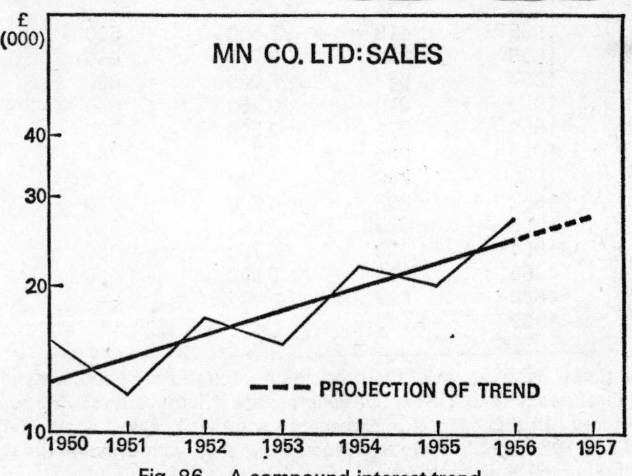

Fig. 86.—A compound interest trend.

Separation of the trend. Among the methods possible are the following:

1. *Curve by inspection.* Plot the series and draw a smooth curve by inspection in such a way that it lies between the points, so that the fluctuations in one direction are equal to those in the other direction, and show the general movement.
2. *The moving average method.* This is the easiest method, but it has a number of disadvantages. It does not cover the complete period and usually it will not represent the trend quite faithfully.
3. *The method of least squares.* A graph of the series will be required to enable a decision to be made as to the particular curve to be fitted.

The calculation of moving averages. The number of items taken for each average will be the number required to eliminate the fluctuations; it will be the number required to cover the period over which the fluctuations occur. In Example 1, since the fluctuations are quarterly, a four-quarterly moving average is required. If monthly figures had been given, a twelve-monthly moving average would have been necessary. Where the period is not obvious, as in the case of cyclical fluctuations, which are not seasonal fluctuations, it will be necessary to draw a graph; the period taken is that between the "peaks" or, alternatively, between the "depressions" on the graphs,

Example 1.

Year	*Production (tons)*	*5-Yearly total*	*Trend*
1953	425	—	—
1954	526	—	—
1955	419	2,500	500
1956	487	2,750	550
1957	643	3,000	600
1958	675	3,250	650
1959	776	3,500	700
1960	669	3,750	750
1961	737	4,000	800
1962	893	4,250	850
1963	925	4,500	900
1964	1,026	4,750	950
1965	919	5,000	1,000
1966	987	—	—
1967	1,143	—	—

If a graph were drawn of the production through the period, it would be seen that "peaks" occur every five years; hence a five-yearly moving average is required. In this case, it is possible to see that "peaks" occur in 1954, 1959 and 1964, that is every five years, without actually drawing the graph. In actual practice, the trend would not, of course, be so regular.

The figures are calculated as follows:

Totals: 425 + 526 + 419 + 487 + 643 = 2,500,
526 + 419 + 487 + 643 + 675 = 2,750,
419 + 487 + 643 + 675 + 776 = 3,000,

and so on.

Averages: $\frac{2,500}{5} = 500$, $\frac{3,000}{5} = 600$, and so on.

The first total and moving average is opposite the third item, *i.e.* the middle item. There are no figures for the first two and the last two items. If there had been seven items to average, there would have been no figures for the first three, nor for the last three items, and the first total and average would have been opposite the fourth item. The case of an even number of items will be dealt with later.

Fitting a trend line by the method of least squares. Only the case where the trend is appropriately represented by a straight line will be considered in this book. A graph is the best means of determining whether a straight line trend is appropriate.

Example 2.

Year	*Sales (£000)*	*Time deviation from middle year*	*Deviations squared*	*Product of sales and deviations*
1963	15	−2	4	−30
1964	17	−1	1	−17
1965	21	0	0	0
1966	23	1	1	23
1967	26	2	4	52
	102		10	28

Average sales $= \frac{102}{5} = 20{\cdot}4$. This is the trend point value for 1965, the middle year.

$$\text{The rate of growth} = \frac{\Sigma\ (\text{Product of sales and deviations})}{\Sigma\ (\text{Deviations squared})}$$

$$= \frac{28}{10} = 2{\cdot}8.$$

The trend is computed as follows:

1963 Average sales − 2 × 2·8 = 20·4 − 5·6 = 14·8.
1964 do. − 1 × 2·8 = 20·4 − 2·8 = 17·6.
1965 Average Sales = 20·4.
1966 Average sales + 1 × 2·8 = 20·4 + 2·8 = 23·2.
1967 do. + 2 × 2·8 = 20·4 + 5·6 = 26·0.

[TREND = Average Sales + Rate × Time Dev.]

A further example is given, this time with an even number of years.

Example 3.

Year	Sales	Time deviation	Deviations squared	Product of sales and deviations
1962	12	−2·5	6·25	−30·0
1963	15	−1·5	2·25	−22·5
1964	17	−0·5	0·25	− 8·5
1965	21	+0·5	0·25	+10·5
1966	23	+1·5	2·25	+34·5
1967	26	+2·5	6·25	+65·0
	114		17·50	49·0

Average sales = $\frac{114}{6}$ = 19. This is the trend point value for the middle of the period, *i.e.* half way between 1964 and 1965. Hence 1964 is half a year before the middle point and 1965 is half a year after the middle point.

Sales grow by an amount equal to $\frac{49}{17{\cdot}5}$ = 2·8.

The trend is as follows:

1962	Average sales	− 2·5 × 2·8 = 19 − 7·0 = 12·0.
1963	do.	− 1·5 × 2·8 = 19 − 4·2 = 14·8.
1964	do.	− 0·5 × 2·8 = 19 − 1·4 = 17·6.
1965	do.	+ 0·5 × 2·8 = 19 + 1·4 = 20·4.
1966	do.	+ 1·5 × 2·8 = 19 + 4·2 = 23·2.
1967	do.	+ 2·5 × 2·8 = 19 + 7·0 = 26·0.

When, instead of a constant amount of increase or decrease each year, there is a uniform percentage increase or decrease each year—that is to say, if the data were plotted on semi-logarithmic paper, a straight-line trend would be appropriate—the above method can still be used. In this case, however, logarithms are used as the basis of the calculation, instead of the original data. If it were required to fit such a compound-interest trend to the above data, logarithms of the sales would be substituted for the sales, the calculations would then be carried out in exactly the same way, and the trend would then be calculated in the form of logarithms. These would then be required to be transformed into numbers by means of tables. This method is useful for estimating increases of population as population increases according to the compound interest law.

Seasonal fluctuations—method of averages. This method ignores any trend that may be present. An example shows the necessary calculations.

Example 4.

Sales (£000)

Quarter	*1964*	*1965*	*1966*	*Total*	*Percentage of each quarterly total to average quarterly total*
1	3·6	3·7	4·2	11·5	104·5
2	2·5	3·4	3·0	8·9	80·9
3	3·9	3·7	3·1	10·7	97·3
4	4·6	4·5	3·8	12·9	117·3
				44·0	400·0

A much longer series is required for the estimates of seasonal fluctuations to be valid. Only three years were taken in the example to save long calculations.

The total in respect of each quarter is obtained. These are the figures given in the total column. The average quarterly total is then calculated. In Example 4, this is 44 divided by 4, which gives 11. The percentage of each quarterly total to average quarterly total is then worked out. Thus:

$$\text{Quarter 1 } \frac{11{\cdot}5}{11} \times 100 = 104{\cdot}5.$$

$$\text{Quarter 2 } \frac{8{\cdot}9}{11} \times 100 = 80{\cdot}9,$$

and so on.

If there are not the same number of years for each quarter, the average for each quarter will be calculated, and the average of these averages also calculated. The percentages of each quarterly average to the average quarterly average will then be computed. These will be the seasonal fluctuations.

Example 5.

Quarter	*Total*	*Quarterly average*	*Percentage of each quarterly average to average quarterly average*
1	11·5	$\frac{11{\cdot}5}{3} = 3{\cdot}83$	$\frac{3{\cdot}83}{3{\cdot}66} \times 100 = 104{\cdot}5$
2	8·9	$\frac{8{\cdot}9}{3} = 2{\cdot}96$	$\frac{2{\cdot}96}{3{\cdot}66} \times 100 = 80{\cdot}9$
3	10·7	$\frac{10{\cdot}7}{3} = 3{\cdot}56$	$\frac{3{\cdot}56}{3{\cdot}66} \times 100 = 97{\cdot}3$
4	12·9	$\frac{12{\cdot}9}{3} = 4{\cdot}30$	$\frac{4{\cdot}30}{3{\cdot}66} \times 100 = 117{\cdot}3$

Average quarterly average = 14·65 ÷ 4 = 3·66.

Had there not been the same number of years for each quarter, the totals would not all have been divided by the same number, *e.g.* 3. But, having obtained all the quarterly averages, the other calculations would have been precisely similar.

In the case of a monthly series, the procedure would be exactly the same; the total of the seasonal indices would, however, add up to 1,200 instead of 400.

Seasonal fluctuations—method of moving averages. This method involves finding the trend, taking the trend from the original data, and averaging the resultant fluctuations.

Example 6. The data are from the table on page 169. A four-quarterly moving average is required to find the trend. This means averaging four items, so there will be no trend figure for the first two, nor for the last two items. Since a straightforward moving average would place the averages between the quarters, it is necessary to average adjacent pairs of the moving averages so that the trend values shall correspond to the same periods of the original data. This is done as follows:

Production (000 tons)	*4-Quarterly total*	*Add in pairs*	*Trend (previous column divided by 8)*
83	—	—	—
81	—	—	—
	376		
98		793	99
	417		
114		866	108
	449		
124		915	114
	466		
113		970	121
	504		
115	543	1,047	131
152	592	1,135	142
163	645	1,237	155
162	668	1,313	164
168	696	1,364	171
175	714	1,410	176
191	730	1,444	181
180	752	1,482	185
184	—	—	—
197	—	—	—

The four-quarterly figures are obtained as follows:

$$83 + 81 + 98 + 114 = 376,$$
$$81 + 98 + 114 + 124 = 417,$$
$$98 + 114 + 124 + 113 = 449,$$

and so on.

These totals, shown in the second column, are placed between the periods, since they relate to the mid-point.

The "add in pairs" figures are obtained as follows:

$$376 + 417 = 793,$$
$$417 + 449 = 866,$$
$$449 + 466 = 915,$$

and so on.

The first total is placed opposite the third quarter, the next total opposite the next quarter, and so on.

The "sums in pairs" are divided by 8. The resultant figures give the trend. They are shown in column 4.

Production

Production		Trend	Fluctuations (production less trend)	Percentage production to trend
1968 3rd Qr.	98	99	−1	99·0
4th Qr.	114	108	+6	105·5
1969 1st Qr.	124	114	+10	108·8
2nd Qr.	113	121	−8	93·4
3rd Qr.	115	131	−16	87·8
4th Qr.	152	142	+10	107·0
1970 1st Qr.	163	155	+8	105·1
2nd Qr.	162	164	−2	98·8
3rd Qr.	168	171	−3	98·2
4th Qr.	175	176	−1	99·4
1971 1st Qr.	191	181	+10	105·5
2nd Qr.	180	185	−5	97·3

Fig. 87.—Two methods of measuring fluctuations.

In the case of monthly figures, a twelve-monthly moving total would be required. They would then be summed in pairs and the sums divided by 24. The first six items and the last six would have no corresponding trend value.

The fluctuations from the trend are now averaged to give the seasonal fluctuations. These fluctuations can be regarded as the difference between the trend and the original data or, alternatively, production as a percentage of the trend. This latter view is more reasonable if the trend is changing rapidly. Fig. 87 shows the necessary calculations.

The next stage in obtaining the seasonal fluctuations is to average the fluctuations as computed in the manner just shown. These fluctuations include random fluctuations as well as the seasonal fluctuations. Averaging has the effect of eliminating random fluctuations.

If the seasonal fluctuations are considered as the differences between the original data and the trend, they will be computed as follows:

	1st Qr.	2nd Qr.	3rd Qr.	4th Qr.
1968	—	—	− 1	+ 6
1969	+10	− 8	−16	+10
1970	+ 8	− 2	− 3	− 1
1971	+10	− 5	—	—
	+28	−15	−20	+15
Unadjusted average	+ 9·3	− 5·0	− 6·7	+ 5 = + 2·6
Adjustment	− 0·65	− 0·65	− 0·65	− 0·65 = − 2·6
Seasonal fluctuations	+ 8·65	− 5·65	− 7·35	+ 4·35 = 0

Fig. 88.—Computation of seasonal fluctuations.

The fluctuations for each quarter are averaged. These fluctuations should add up to zero. As will be noted, in the example given they add up to 2·6. The adjustment consists in subtracting one quarter of this difference from each quarter, thus making the adjusted averages add up to zero. These adjusted averages are the seasonal fluctuations.

If the seasonal fluctuations are considered as a percentage of the trend, the seasonal indices, as they are then called, are computed as follows:

	1st Qr.	2nd Qr.	3rd Qr.	4th Qr.
1968	—	—	99·0	105·5
1969	108·8	93·4	87·8	107·0
1970	105·1	98·8	98·2	99·4
1971	105·5	97·3	—	—
	319·4	289·5	285·0	311·9
Unadjusted average	106·5	96·5	95·0	104·0 = 402·0
Seasonal index	105·9	96·0	94·6	103·5 = 400·0

Fig. 89.—Computation of seasonal indices.

The fluctuations for each quarter are averaged. These should add up to 400. As will be noted, in Fig. 89 they add up to 402. The adjusted averages are obtained by multiplying each average by 400 and dividing by 402, thus making the adjusted averages add up to 400.

In the case of monthly seasonal indices they will add up to 1,200.

Seasonal indices are a more logical way of dealing with seasonal fluctuations.

Deseasonalised data. If the seasonal fluctuations are deducted from the original data, the result is the deseasonalised data. This will consist of the trend and the residual fluctuations. Alternatively, the original data is divided by the seasonal indices to give the deseasonalised data. An example of the first method, using the same data as before, is shown in Fig. 90.

Production

	Pro-duction	Seasonal fluctua-tions	Deseason-alised data	Trend	Residuals
1968 3rd Qr.	98	−7·4	105·4	99	6·4
4th Qr.	114	+4·4	109·6	108	1·6
1969 1st Qr.	124	+8·6	115·4	114	1·4
2nd Qr.	113	−5·6	118·6	121	−2·4
3rd Qr.	115	−7·4	122·4	131	−8·6

Fig. 90.—Analysis of a time series.

Testing the suitability of a seasonal index. A separate graph is required for each quarter or month, as the case may be. The years are plotted along the "horizontal" axis. The percentages of the moving averages or the fluctuations (original data less trend) are plotted along the "vertical" axis. Any trends show that the method of computation of the seasonal indices or seasonal fluctuations is not suitable; changing rather than stable indices are required.

Simple forecasting. The figures of production of Amigo Cement Co. Ltd. are given up to and including the fourth quarter of 1971. It is required to forecast the production for the first two quarters of 1972. The procedure is as follows:

1. Find the trend value for the first two quarters of 1972.
2. Add the appropriate seasonal variations or, better still, multiply by the appropriate seasonal indices.

A simple way of finding the trend values is shown in Fig. 91.

	Trend		*Trend*
3rd Qr. 1969	131	3rd Qr. 1970	171
4th Qr. 1969	142	4th Qr. 1970	176
1st Qr. 1970	155	1st Qr. 1971	181
2nd Qr. 1970	164	2nd Qr. 1971	185
	592		713

Mean value of trend $\frac{592}{4} = 148$ $\qquad \frac{713}{4} = 178$

Increase in mean value over one year = 178 − 148 = 30

Increase in mean value per quarter $= \frac{30}{4} = 7{\cdot}5$

	Trend
2nd Qr. 1971	185
3rd Qr. 1971 = 185 + 7·5	= 192·5
4th Qr. 1971 = 192·5 + 7·5	= 200
1st Qr. 1972 = 200 + 7·5	= 207·5
2nd Qr. 1972 = 207·5 + 7·5	= 215

Fig. 91.—Compilation of future trend values.

The forecasts for the first two quarters of 1972 are:

$$\text{1st Qr. 1972 } 207{\cdot}5 \times \frac{105{\cdot}9}{100} = 220 \text{ thousand tons.}$$

$$\text{2nd Qr. 1972 } 215 \times \frac{96}{100} = 206 \text{ thousand tons.}$$

Exponential smoothing. This is a method of short-term forecasting. The only information needed is (1) the forecast for the current period—the old forecast—and (2) the value for the current period—the observation. It is then possible to make a forecast for the next period—the new forecast.

The new forecast is calculated as follows:

NEW FORECAST = OLD FORECAST + α (OBSERVATION − OLD FORECAST)

where α is the *smoothing factor* or *smoothing constant.*

If $\alpha = 0$, then the new forecast = the old forecast, that is, there is complete stability; if $\alpha = 1$, then the new forecast = latest observation, that is, there is complete sensitivity. In most cases a smoothing factor between 0·1 and 0·2 will be found suitable. The smaller the value of the smoothing factor, the smoother the forecast series, that is, the smaller the deviations.

The monthly production of a firm is as follows: January 38,000 tons, February 34,000, March 40,000, April 36,000, May 32,000, and June 38,000. Fig. 92 shows the calculations of the forecasts, and Fig. 93 enables a comparison to be made between the actual production and the forecasts.

Monthly Production (thousand tons)

	Old forecast	*Observation*	*Error (observation − old forecast)*	α *(error)* (α = 0·2)	*New forecast*
	(*a*)	(*b*)	(*c*) (*b*) − (*a*)	(*d*) α(*c*)	(*e*) (*d*) + (*a*)
Jan.	32·0	38	6·0	1·2	33·2
Feb.	33·2	34	0·8	0·2	33·4
Mar.	33·4	40	6·6	1·3	34·7
Apr.	34·7	36	1·3	0·3	35·0
May	35·0	32	−3·0	−0·6	34·4
June	34·4	38	3·6	0·7	35·1

Fig. 92.—Calculation using exponential smoothing.

In the example just given there is no trend. Where a trend exists a more complicated formula is required. To the formula for exponential smoothing of the observations are added terms to provide for the exponential smoothing of the trend.

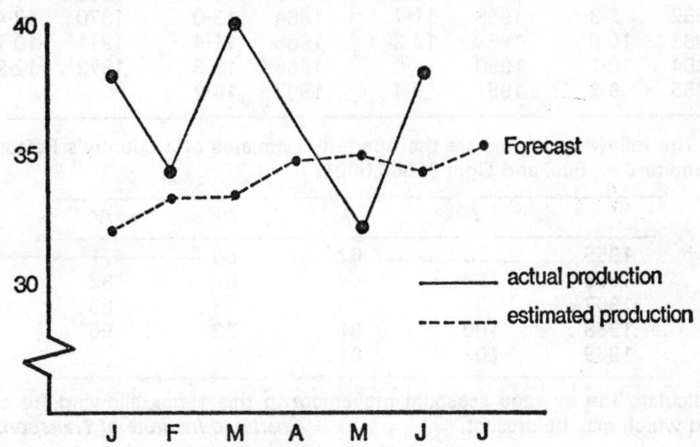

Fig. 93.—Exponential smoothing of a time series.

When the data require adjustment before analysis. Over the past thirty years the value of money has continuously changed. Any time series expressed in terms of money will show these changes in addition to the actual changes in the data. In most cases the data will require to be deflated by a suitable price index, often specially computed for the purpose, before analysing the time series, in order to eliminate this effect of the changing value of money.

In the case of monthly figures, adjustments may be necessary because of the varying length of the month. The original data is multiplied by an appropriate correction factor.

Examples of such factors are:

(*a*) $\dfrac{\text{average number of days per month}}{\text{actual number of days in the month}}$.

(*b*) $\dfrac{\text{average number of working days per month}}{\text{actual number of working days in the month}}$.

Questions

1. From the data below compute the trend, stating your method of determining the period of the cyclical fluctuations.

Year		*Year*		*Year*		*Year*	
1950	11·1	1956	11·4	1962	10·0	1968	12·6
1951	13·1	1957	11·5	1963	10·3	1969	10·1
1952	8·3	1958	11·7	1964	13·0	1970	12·4
1953	10·0	1959	12·3	1965	11·4	1971	10·7
1954	10·1	1960	11·0	1966	12·3	1972	12·9
1955	9·2	1961	9·4	1967	14·2		

2. The following figures are the quarterly estimates of a country's Personal Expenditure on Fuel and Light (£000,000):

	I	*II*	*III*	*IV*
1965	78	62	56	71
1966	84	64	61	82
1967	92	70	63	85
1968	100	81	72	96
1969	107	81		

Calculate the average seasonal movement in this series allowing for any trend which may be present. *Chartered Institute of Transport.*

3.

Personal Expenditure on Fuel and Light
£(*million*)

Quarters	*1*	*2*	*3*	*4*
1969	105	77	68	95
1970	107	83	74	106
1971	117	99	86	112

Explain how seasonal variations are removed from time series.
Graph the series above and insert the trend of expenditure.

4. Write brief notes on:

(*a*) class interval
(*b*) compensating error
(*c*) primary date
(*d*) smoothing of curves

Institute of Cost and Management Accountants.

5. The following table shows the amount of money (£000,000) spent upon passenger travel in Barbalonia at levels of fares, and charges current, during the periods given:

	1963	*1964*	*1965*	*1966*	*1967*	*1968*
Quarter I	71	71	73	76	84	89
II	89	90	91	97	106	110
III	016	108	111	122	130	—
IV	78	79	81	89	95	—

Calculate the seasonal pattern for this series and estimate the values of the series for the third and fourth quarters of 1968.

6. The following table shows the amount of money (£000,000) spent upon passenger travel in Barbalonia for the periods given when valued at the level of fares and charges in force during 1963:

	1963	1964	1965	1966	1967	1968
Quarter I	71	71	73	71	74	76
II	89	88	91	89	91	94
III	106	104	110	113	112	—
IV	78	78	80	81	81	—

Using this series and the figures in Question 5, plot a scatter chart. What general information does this afford on the question of changes in the levels of fares and charges?

7. *Saleable Output of Open-cast Coal (in thousands of tons per quarter)*

Year	*I*	*II*	*III*	*IV*
1965	—	—	—	2,595
1966	2,340	3,051	3,061	2,565
1967	2,458	3,161	3,466	2,793
1968	2,809	2,911	3,035	2,917
1969	2,652	2,776	2,634	—

From the above table:

(*i*) compute the mean seasonal variation present;
(*ii*) rewrite the figures for the year 1969 with the seasonal variation removed.

8. Coal-Mining Industry
Number of Newly Employed Juveniles under Eighteen Years

	Quarters			
	1	*2*	*3*	*4*
1964	3,934	3,792	3,860	2,569
1965	3,872	3,823	4,522	4,902
1966	7,178	3,503	5,411	3,493

Rewrite these numbers to the nearest hundred and then calculate the trend and seasonal variation.

9. Assuming the true trend of the number of trucks and wagons owned by the main line railway companies was represented by the years 1950 to 1959—estimate the figure for 1965:

End 1950	634	End 1958	683
1950	634	1959	687
1952	657	1960	678
1953	664	1961	665
1954	664	1962	656
1955	670	1963	689
1956	672	1964	691
1957	678	1965	(?)

10. *Total Production of Paper*
Weekly averages (thousand tons)

Year	*Quarters*			
	1	*2*	*3*	*4*
1966	37	38	37	40
1967	41	34	25	31
1968	35	37	35	41

Calculate the moving average trend of the above data and estimate the seasonal variation in the weekly paper production.

Draw on the same diagram the weekly average production together with the trend.

11. The production of small cigars in the United States 1961–70 is shown in the following table.

Year	*Number of small cigars (millions)*
1961	98·2
1962	92·3
1963	80·0
1964	89·1
1965	83·5
1966	68·9
1967	69·2
1968	67·1
1969	58·3
1970	61·2

You are required to:

(*a*) graph the data shown above,

(*b*) calculate the equation of a least squares trend line fitting the data, say what the trend value is at 1961 and 1970, and explain your result, and

(*c*) estimate the production of small cigars for the year 1971.

Association of Certified Accountants.

12. Smooth exponentially the following time series and forecast the profits for August 1976. Use a smoothing constant of 0·1.

The profits of Ansco were: January 1976 £52,000, February £64,000, March £56,000, April £54,000, May £58,000, June £60,000, July £54,000.

(*Note:* At the start of a series any reasonable value may be taken as the "old forecast". Let it be £52,000 in this case.)

13. Quarterly production for a paper-making plant was reported as follows:

Years	*Quarterly production ('00 tonnes)*			
	I	*II*	*III*	*IV*
1972	38·8	41·3	39·0	45·6
1973	44·7	45·2	42·0	49·9
1974	46·7	48·2	44·5	51·3
1975	50·1	54·6	—	—

Using the method of moving averages, find the average seasonal deviations and thus estimate the production figures for the last two quarters of 1975.

I.C.S.A. Part 1, Dec. 1975.

14. Required:

In the analysis of time series

(1) explain what is meant by a trend line and describe very briefly two principal methods of calculating it;

(2) explain what is meant by residual variations and why they are important;

(3) give two situations where it is important to adjust the data prior to analysis, with reasons;

(4) sketch a graph of a time series which shows a trend and also cyclical and seasonal movements.

Association of Certified Accountants.

CHAPTER 26

STATISTICAL QUALITY CONTROL

Controlling quality by statistical methods. Statistical quality control is a method of estimating the quality of the whole from the quality of the samples taken from the whole. The method is based upon the laws of chance and has a sound mathematical basis.

Among the many advantages of statistical quality control, the most important is that it is more efficient. 100 per cent inspection (*i.e.* inspecting every article) is not reliable; the inspector usually inspects as many as possible and passes the rest. Human fatigue also causes errors of inspection. Quality control charts, on the other hand, indicate not only whether the quality is actually up to the required standard, but, as will be seen later on in the chapter, give warning of possible future lack of quality which can hence be avoided. Another important advantage may be a saving in cost. Since only a part of the whole is inspected, fewer inspectors are needed; there is a saving in time and wages.

Theoretical foundation of quality control. A machine may be set to cut out a hole of a given diameter in, say, a piece of metal. Yet although there is no cause, such as a fault developing in the machine or the operator becoming tired or careless, the measurements will vary. For this reason tolerance limits are fixed. These are the largest and smallest sizes between which the measurement must fall to satisfy the required specification. When there is no assignable cause, *i.e.* a specific reason for variations in measurements such as those just mentioned, the measurements of the pieces will vary according to the "normal" distribution.

This "normal" distribution has the following characteristics:

(*a*) 998 out of every 1,000 pieces will vary in size between the average size plus 3·09 times the standard deviation and the average size less 3·09 times the standard deviation. This is expressed in symbols as $\bar{X} \pm 3{\cdot}09\ \sigma$ (pronounced bar X plus or minus 3·09 sigma), $\bar{X}$ standing for the average size and σ standing for the standard deviation.

(*b*) 38 out of every 40 pieces will vary between the average size plus 1·96 times the standard deviation and the average size less 1·96 times the standard deviation.

This distribution occurs when pure chance alone causes variations in the sizes. If, therefore, more than one piece in a thousand were larger than $\bar{X} + 3{\cdot}09\ \sigma$, it would be necessary to find an assignable cause, one that can be removed. Variations due to pure chance cannot be removed.

The entirety of the batch is known as the population. The average size of the population is denoted by $\bar{X}$ and the standard deviation by σ_p (the "$_p$" denoting population).

However, the only information available will be from the samples taken. No information can be obtained from a single sample, although in practice a population is often judged by a single sample, which is most unscientific and completely unreliable. The average measurement of the samples is considered the best estimate of the average measurement of the population. The greater the number of samples, the better the estimate.

The measurement of each piece in each sample is taken. In respect of each sample it will therefore be possible to find the range of measurements of each sample and also the average measurement of each sample. This average measurement of each sample will also vary according to the "normal" distribution (again provided there is no assignable cause why it should not). The averages have, however, as their standard deviation—$\frac{\sigma_p}{\sqrt{n}}$, where n is the number of items in each sample—the number of items in each sample will always be the same.

The variation in the sizes in the individual items and the variation in the average measurements of the samples MUST NOT BE CONFUSED.

The average measurements of the samples, since they vary according to the "normal" distribution, will have as their limits (apart from 2 in every 1,000) $\bar{X} \pm \frac{3{\cdot}09\ \sigma_p}{\sqrt{n}}$. These are known as the control limits for means.

$\frac{3{\cdot}09\ \sigma_p}{\sqrt{n}}$ is calculated by multiplying the average range of the samples, denoted by $\bar{w}$, by an A factor. This factor varies according to the number in the sample. If the sample averages fall outside the limits, the individual items' measurements will also fall outside their limits. It will also be necessary to look for an assignable cause.

The ranges of all the samples might vary considerably, and yet the average range still be within the permissible limit. The calculation of the variation of the sample measurements is based on the average range. It is therefore necessary to plot the ranges of the individual samples on a range chart and verify that they do not fall outside the control limit of ranges. This control limit is arrived at by multiplying the average range by a D factor. This factor also varies according to the number of items in the sample. In the case of ranges, only the upper limit is of relevance. In actual practice it has been found unnecessary to continue plotting the range chart when it has been found satisfactory for a comparatively short period.

Control charts. The means control chart enables possible trouble to be averted. The principal indications when action should be taken are:

(*a*) an average outside the control limits,
(*b*) several averages, especially if consecutive, near a control limit,
(*c*) an undue number of averages above or below a central value,
(*d*) a trend in the averages,
(*e*) an unusually long run of averages above or below a central value.

The first step in the making of control charts is to collect the data and enter the data on sheets. Measurements are taken in respect of each component in each sample. These are entered on sheets, together with details of tolerance limits, date and time of samples, operator's number, the inspector's name, details of the component and the dimension being controlled, and any other relevant details. Samples are taken at suitable convenient periods.

The measurements in respect of each sample will be added and divided by the number in the sample. This will give the average measurement of the sample. The largest measurement in each sample less the smallest measurement will give the sample range. These figures will also be entered on the data sheet.

The next step is to plot these sample means and ranges on two separate graphs both drawn on the same sheet of paper. The scales should be carefully chosen to suit the particular measurements of the component's dimension being dealt with. The scale is shown on the "vertical" axis. The tolerance limits should also be marked on the means chart.

The third stage is to insert temporary control limits. It has been suggested (*A First Guide to Quality Control for Engineers*, H.M.S.O.) that temporary control limits be fixed after ten samples have been taken. This is done as follows:

(*a*) The ten sample means are averaged to find the grand mean, denoted by $\bar{X}$.
(*b*) The ten sample ranges are averaged to give average range, denoted by $\bar{w}$.
(*c*) $A\bar{w}$ and $D\bar{w}$ are computed. The values of A and D will be found in Appendix F.
(*d*) Draw horizontal lines on the means chart at (1) $\bar{X}$, (2) $\bar{X} + A\bar{w}$, (3) $\bar{X} - A\bar{w}$. Lines (2) and (3) are the upper and lower control limits respectively.

(*e*) Draw horizontal lines at $\bar{w}$ and $D\bar{w}$. The latter line is the control limit for ranges. Since these control limits are temporary, they should be drawn in pencil.

The final stage is to compute new control limits after a further 15 samples have been taken, making 25 in all. The whole of the data of the 25 samples should be used. These final limits should be entered in red ink, the averages being in blue.

Once controll imits have been established, no point should be allowed to fall outside without investigating the reason. When points no longer fall outside, the process is said to be under control.

Statistical control and tolerance limits. A machine may be turning out parts whose variations in size are entirely due to pure chance. The process is under statistical control. Nevertheless the variations may be greater than those necessary to comply with the specification. *It is therefore important to make certain that the variation in the sizes of the individual items is not greater than the variations allowed by the tolerance limits.*

The tolerance range—the difference between the largest size and the smallest allowed by the specification—is multiplied by an L factor, which varies with the number of items in the sample. Provided the average range of the samples does not exceed this figure, the variation of the individual items coming from the machine is within the variation allowed by the specification. If, however, $\bar{w}$ exceeds L times the tolerance range, the control chart will be unsatisfactory; $\bar{w}$ MUST NOT EXCEED $L(T_u - T_u)$. The alternatives are 100 per cent inspection or the purchase of a more precise machine.

One further test remains before the control chart can be considered satisfactory. The amount of variation may be satisfactory, but it must also be between the specified measurements. To test this it is necessary to verify that the control limits lie between $T_u - M\bar{w}$ and $T_L + M\bar{w}$, where T_u is the upper tolerance limit, and T_L is the lower tolerance limit, and M is a factor which varies according to the number of items in the sample. If the control limits do not lie within this band, but the tolerance range is satisfactory, the machine will require resetting.

If the control limits are within the allowable width, but are widely separated from them, the machine being used for the process is too precise. A cheaper and less precise machine might be used.

Means charts with two sets of control limits. The 3·09 sigma limits or 1/1,000 limits are the more usual. Sometimes, however, 1/40 limits are used in addition. When a reading falls outside these latter limits it is taken as a warning, and an investigation takes place. When, however, a reading falls outside the former, it is considered that there is definitely something wrong.

Worked example. The internal diameter of a ring is required to be 3 mm, with a tolerance of 0·24 mm. Samples of 4 are taken every half-hour. The first samples were found to have a mean measurement of 3·01 mm and an average range of 0·12 mm. The values of the

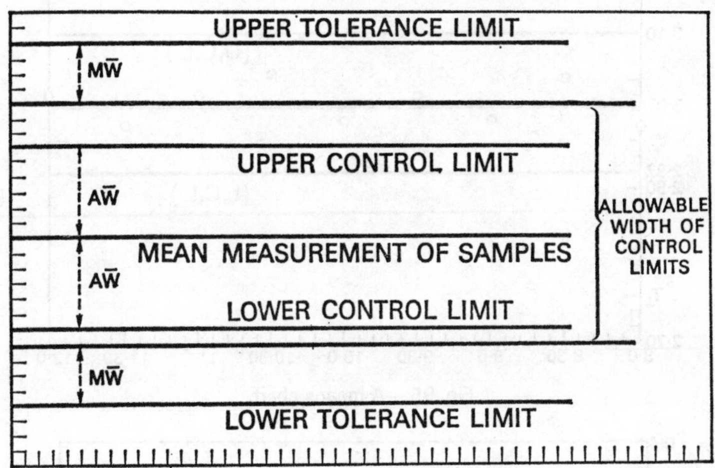

Fig. 94.—The relationship between control limits, tolerance limits and allowable width of control limits.

factors *A*, *L*, *M* and *D* for samples of 4 are 0·75, 0·33, 0·75 and 2·57 respectively. Draw a means chart and a ranges chart and verify if they are satisfactory.

1. *Calculation of control limits for means.*
   Upper control limit: $3{\cdot}01 + 0{\cdot}75 \times 0{\cdot}12 = 3{\cdot}10$ mm.
   Lower control limit: $3{\cdot}01 - 0{\cdot}75 \times 0{\cdot}12 = 2{\cdot}92$ mm.
2. *Verification that variation of sizes of items is not greater than tolerance spread.*
   Difference between greatest and smallest size allowable 0·48 mm. $L \times 0{\cdot}48 = 0{\cdot}33 \times 0{\cdot}48 = 0{\cdot}16$. Since 0·12 does not exceed 0·16 the variation of individual sizes is satisfactory.
3. *Calculation of allowable width of control limits.*
   Upper tolerance limit less $M\bar{w} = 3{\cdot}24 - 0{\cdot}75 \times 0{\cdot}12 = 3{\cdot}15$.
   Lower tolerance limit plus $M\bar{w} = 2{\cdot}76 + 0{\cdot}75 \times 0{\cdot}12 = 2{\cdot}85$.
   Control limits for means are within allowable widths.
4. *Calculation of control limit for ranges.*
   Control limit. $D\bar{w} = 2{\cdot}57 \times 0{\cdot}12 = 0{\cdot}30$.

On the charts shown in Figs. 95 and 96 measurements are plotted relating to the period subsequent to the taking of the 25 samples.

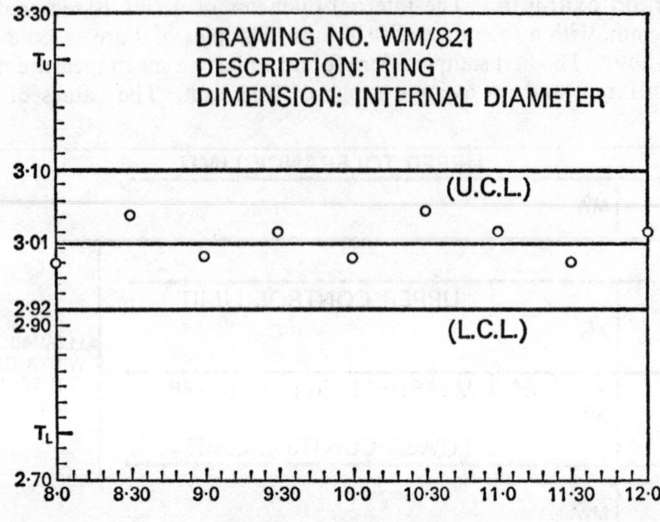

Fig. 95.—A means chart.

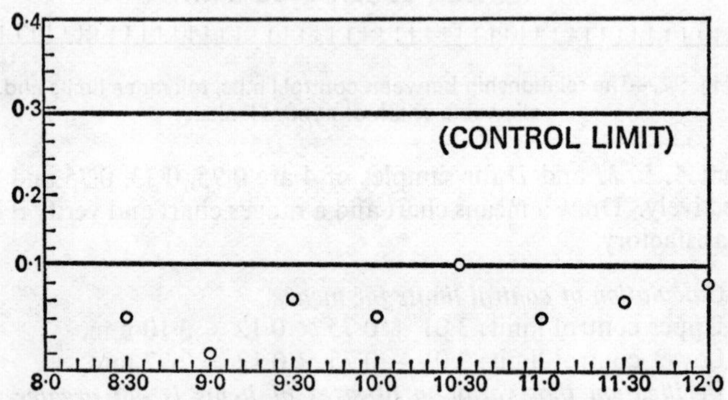

Fig. 96.—A ranges chart.

Questions

For factors refer to Appendix F.

1. What do you know of "Statistical Quality Control"? Give some account of its basis and methods. *Institute of Company Accountants.*

2. A piece of metal is required 10·75 inches long. The tolerance allowable is 0·15 inches either way. The average length of a large number of samples was 10·76 inches. The average range was 0·08 inches. Each sample consisted of 5 pieces. You are required to draw a means control chart and to verify if it is satisfactory.

3. What advantages does statistical quality control possess over 100 per cent inspection of production or sampling which is not based on statistical theory? *Institute of Cost and Management Accountants.*

4. At a factory pink and white sweets are produced in separate plants. Approximately equal streams of each colour are fed into a single large mixer, where the sweets are well mixed. Batches, each of 100 sweets, are drawn from the mixer and a count is made of the number of sweets of each colour. 400 batches were examined and the following information was obtained:

Average number of pink sweets, per batch	50
Standard deviation of the number of pink sweets per batch	5

(*a*) What is the standard deviation of the number of white sweets per batch?

(*b*) What is the maximum number of pink sweets that you would expect to find in any one batch taken at random?

(*c*) If the sweets were drawn off in batches of 400 what would you expect the average number of pink sweets per batch (of 400) to be, and what would you expect the standard deviation of the number of pink sweets per batch to be?

(*d*) For batches of 100, draw a control chart on which the results of the colour counts could be entered. Enter on this chart points which would indicate that the mixing was at first satisfactory, but had then become unsatisfactory. Give your reasons in each case. *Institute of Cost and Management Accountants.*

5. The interior diameter of a ring is required to be 120 mm with an allowable tolerance of 0·25 mm either way. The average diameter of a large number of samples was 119·97 mm and the average range was 0·13 mm. Each sample consisted of 6 rings.

You are required to:

(*a*) draw a means control chart showing both action and warning limits and say whether the chart is satisfactory.

(*b*) draw a ranges chart.

CHAPTER 27

DEMOGRAPHY

Population statistics. The subject-matter of demography, these are of great practical value in planning ahead (future demand for housing, schools, hospitals, energy and future supply of labour for manpower planning) as well as being of value in the present, for among other data, they give sex, age and area distributions, basic data for estimating demand of commodities and services and supply of labour.

Population. The basis of all population figures in this country is the decennial census (there was no census in 1941). It is a "*de facto*" census, *i.e.* it is a census of the actual population, of people in the country at a certain date. This is very carefully defined. It includes

> "all persons enumerated at census midnight on land, in barges and boats on inland waters, in all vessels (other than ships of foreign navies) in port or at anchorages, also persons in vessels on fishing or coastwise voyages which returned to port during April, not having proceeded from a port outside Great Britain, Ireland or their adjacent islands. The population does not include members of the armed forces, mercantile marine or other civilians *outside* the country but does include those units of the armed forces of other countries stationed in the United Kingdom."

Vital statistics. The statistics of births, marriages and deaths depend upon the compulsory registration of the following information.

1. *Births.* When and where born, name, sex, name and surname of father, name and maiden surname of mother, rank or profession of father, signature, description and residence of informant, when registered, baptismal name if added after registration, and in case of still-births, evidence upon which birth is registered.

2. *Deaths.* When and where died, name and surname, sex, age, rank or profession, cause of death, signature, description and residence of informant, when registered.

3. *Marriages.* When married, names and surnames, ages, marital conditions, ranks or professions, residences at time of marriage, fathers' names and surnames, ranks or professions of fathers.

All the above information is entered in public registers. Under the

Population (Statistics) Act, 1938, the following additional confidential information is collected, but not published in any public register:

4. *Births.* Age of mother, date of marriage, number of children by present husband, and how many still living, number of children by any former husband, and how many still living.

5. *Deaths.* In the case of a man, whether he had ever been married, and if married at date of death. In the case of a married woman, the year when married, duration of marriage, and whether she had had any children by her husband. Age of surviving spouse, if any.

Main publications of the Registrar-General. These are as follows:

1. *Weekly return.* This is published on the Saturday following the week to which it relates. It contains, *inter alia*, the number of births, still-births, deaths, infant deaths and deaths from some of the main epidemic diseases registered in the preceding week in the county boroughs and the "Great Towns" of more than 50,000 population (160 in all), a record of deaths by age and principal causes in London, numbers of cases of infectious disease notified in each local government area.

2. *Quarterly return.* This is usually published before the end of the quarter following the one to which it relates. It contains, *inter alia*, estimates of population by sex and certain age groups; a serial record of quarterly and annual figures of births, still-births, marriages, deaths, infant deaths and deaths under 4 weeks; a serial record covering preceding five quarters of birth and death rates; number of deaths from principal causes, distinguishing sex; the number of insured persons absent from work owing to certified sickness or industrial injury on specific days in three consecutive months in England and Wales and in Standard Regions (supplied by the Department of Health and Social Security).

3. *Annual Statistical Review.* This appears in three volumes.

Part I. Medical, dealing with deaths and death-rates; disease notifications.

Part II. Civil, dealing with the statistics of births, marriages and population, and fertility; the numbers of parliamentary and local government electors (from local registers).

Part III. Text, giving a critical commentary on figures published in volumes of tables, together with special analyses and tabulations. An analysis of major trends in vital statistics over the years is given. The "text" volumes have not been published annually in recent years and the subject matter has covered several volumes.

4. *Decennial Supplements.* These are usually published in several

parts. They give analyses of mortality and fertility by occupations, life tables, multiple or secondary causes of death.

5. *Annual Estimates of the Population of England and Wales.* These give annual estimates of the populations of each local government area in England and Wales.

The Registrar-General for Scotland also publishes a weekly return, a quarterly return and an annual report.

The 1971 census. This was the 17th decennial census of England and Wales. Censuses were also taken on the same date, the 25th April, by the appropriate authorities in Scotland, Northern Ireland, the Isle of Man and in Guernsey and the adjacent islands and in Jersey on the 4th April.

Returns were made for every person alive in the country at midnight ending on census day, and certain other persons who, although not in the country at that time, are usually resident in the country, *e.g.* persons engaged in fishing or on a coastwise voyage.

The Registrar-General for England and Wales and the Registrar-General for Scotland are, under the control of the Secretary of State for the Social Services, responsible for the carrying out of the census.

Census field organisation. The country was divided into small areas, known as *enumeration districts.* An enumerator was appointed to each of these districts. He was responsible for enumerating all persons present or resident within his district on census night. There were some 97,000 enumerators.

The enumeration districts form part of larger areas known as *census districts*, each of which is the responsibility of a *Census Officer*, who will normally be the registrar of births and deaths. There were some 2,000 Census Officers. There were also *assistant Census Officers* – one for every sixteen or seventeen enumerators, appointed to control the quality of the enumerators' work and to carry out checking duties on behalf of Census Officers.

There were also appointed 100 supervisors, each covering an area of about a half a million people.

Duties of the enumerators.

(*a*) To deliver and collect census forms for every household and communal establishment.
(*b*) To keep a record book entering details of all buildings.
(*c*) To check forms after collection; to complete a permanent record book.
(*d*) To prepare the population summary and enter some of the material collected on computer input sheets.

Duties of the Census Officer. These are to take charge of the enumeration in his district, to appoint, instruct and to supervise the enumerators. He acts for the Registrar-General in all matters relating to the census in his district. He will receive from the enumerators the Schedules and the completed enumeration records, which, after examination, and completion and correction where necessary, he will forward to the central Census Officer.

Census questions.

1. *Questions A1–A5: household's accommodation.* Type of tenure of accommodation; number of living rooms; whether any room, hall, passage or staircase is shared with anyone else; whether household has hot water supply, bath or shower, sink, cooking-stove, flush toilet; number of cars and vans available.

2. *Questions B1–B24: particulars of every person present in the household.* Name; date of birth; sex; usual address; relationship to head of household; marital condition; whether employed during last week *i.e.* the week ended 24th April; if a student attending full-time at an educational establishment during the term starting April/May 1971; place of birth; country of birth of father and mother; usual address a year ago; usual address 5 years ago; education (G.C.E. "A" Levels, degrees, professional qualifications, etc.); employer's name and business; occupation; whether employee or self-employed and whether employing others; if apprentice or trainee; number of hours a week on job; address of place of work; means of transport; occupation one year ago; (for women under 60 who are married, widowed or divorced) month and year of each child born alive and month and year of marriage.

3. *Questions C1–C7: particulars of persons who usually live in household but who are not present.* Name and relationship to head of household; sex and date of birth; marital status; whether employed last week; name and business of employer; occupation; if employee or self-employed and whether employing others.

Census reports.

1. *Preliminary report.* This contains provisional figures of the population for local authority areas, prepared from summaries sent in by Census Officers.

The population of the U.K. was given as:

Males	26,955,529
Females	28,566,005
Total	55,521,534

2. *County and national reports.* These are based on the computer

analysis of information coded and punched at the central census office directly from the original census forms.

For some census topics, such as details of employment and educational qualifications, the reports will initially be based on a 10 per cent sample of households (or persons in communal establishments).

Questions

1. What is meant by a *"de facto"* census? Give a definition of "actual population of people in this country at a certain date."

2. Give the main publications of the Registrar-General and say what information they contain.

3. How was the 1971 census organised?

4. What information concerning vital statistics must be registered?

5. Outline the procedures which are employed when a Census of Population is undertaken in the United Kingdom, or in a country of your choice.

I.C.S.A. Part 1, Dec. 1975.

CHAPTER 28

ECONOMIC STATISTICS

Monthly Digest of Statistics. This is published monthly by the Stationery Office. It is prepared by the C.S.O. (Central Statistical Office) in collaboration with the Statistics Division of Government Departments. The Digest reproduces the principal official statistics. The table below gives a summary of the main contents of the Digest.

The *Annual Abstract of Statistics* covers a wider range of subject-matter, and in many cases the annual figures are given in greater detail than in the *Monthly Digest*.

Subject-matter	*Main analysis*	*Source*
	National Income and Expenditure	
Expenditure on the gross domestic product	Consumers' expenditure Public authorities' expenditure Gross domestic capital formation Exports and imports of goods and services Taxes on expenditure and subsidies	C.S.O.
Factor incomes in the gross national product	Gross national product. Property income from abroad Income from employment Gross trading profits of companies Gross trading surplus of public corporations Gross profits of other public enterprises Income from rent and self-employment	
	Population and Vital Statistics	
Population	Total for U.K., for England and Wales, for Scotland and for Northern Ireland. Each total is analysed into males and females	Office of Population censuses and Surveys
Age distribution	U.K., England and Wales, Scotland and Northern Ireland, analysed into males and females	General Register Office (Scotland)
Births and marriages registered	Live births and marriages for U.K., England and Wales, Scotland and Northern Ireland. Quarterly figures are given	General Register Office (Northern Ireland)
Deaths registered	Quarterly totals for U.K., England and Wales, Scotland and Northern Ireland Similar figures for infants under one year	

Subject-matter	*Main analyses*	*Source*
	Labour	
Distribution of working population	Total working population analysed into males and females Number in H.M. Forces, in civil employment and registered unemployed	Department of Employment Department of Manpower Services (Northern Ireland)
Employees in employment	Total number of employees, analysed by industry	
Registered unemployed	Monthly figures, males and females unemployed in G.B. and Northern Ireland Wholly unemployed and temporarily stopped Unemployed for Northern Ireland, analysed into males and females Analysis according to regions, duration of unemployment and by industries	
Industrial stoppages	Analysed according to groups of industries	
	Social Services	
National Insurance and family allowances	Claims for sickness benefit and injury benefit, claims for unemployment benefit; pensioners, and family allowances	Department of Health and Social Security
	Industrial Production and Output	
Index of industrial production	For all industries and also for a large number of industries	C.S.O.
Output per head	For all industries and also for a large number of industries	C.S.O. and Department of Employment
	Energy	
Coal	Monthly figures of production, consumption and stocks Wage earners on books Average output in tons per manshift Intake and outflow of manpower	Department of Energy
Gas	Monthly figures of fuel used. Gas sent out Natural gas supply	
Electricity	Generation for public supply, giving amount generated, fuel used and amount sent out	
Petroleum	Arrivals of tanker-borne crude petroleum Refinery throughput and output of finished petroleum products Deliveries of petroleum products for inland consumption	
	Industrial Materials and Manufactured Goods	
A large number of tables relating to chemicals, metals, ship-building, engineering products, electrical goods, vehicles, textiles and manufactured goods	Production, consumption, stocks, in value and quantities, but not in all cases Various analyses according to subject matter	Department of Industry
	Construction	
Permanent house building	For each quarter construction begun, under construction at end of period, completions; giving totals, for local housing authorities and for private owners, and other authorities. These figures are given for U.K., England and Wales, Scotland and Northern Ireland	Department of the Environment Scottish Development Department Department of Housing, Local Government and Planning (Northern Ireland)
Educational building in Great Britain	Projects approved; contracts under construction and contracts completed	Department of Education and Science, Scottish Education Department.

Subject-matter	*Main analyses*	*Source*
Building materials and components	Monthly figures of production and stocks	Department of Industry Department of the Environment
	Agriculture and Food	
A large number of tables relating to all kinds of foods and agricultural products	Production, stocks, and disposals (where appropriate) in quantities	Agriculture Departments
	External Trade	
Value of external trade of U.K.	Values of imports and exports, with a certain amount of analysis	
Indices of volume	Volume index numbers of imports and exports	Department of Trade
Imports into U.K.	Analysis by area Monthly imports of certain commodities (quantities)	
Exports	Monthly quantities and sometimes values of certain commodities Imports by countries Exports by countries	
	Transport	
Passenger journeys	British Rail London Transport Public road transport	
British Rail and inland waterways	Monthly figures of freight traffic	Department of the Environment
Road vehicles	Monthly figures of new registrations Road vehicles in Great Britain	
Road casualties	Analysed according to nature and class of road user	Department of the Environment Scottish Development Department
Merchant shipping	Merchant vessels of 500 gross tons and over registered in the U.K.	Department of Industry
Civil aviation	Miles flown, passengers carried, mail carried, freight carried Monthly figures for all services, domestic, and international services	Civil Aviation Authority
	Finance	
Exchequer revenue and expenditure	Analysis of sources of revenue and categories of expenditure	Treasury Board of Inland Revenue
Bank of England	Gives for each month the average figures of the Weekly Return	
London Clearing Banks	Gives, for one date each month, combined Balance Sheets of London Clearing Banks	Bank of England
Bank advances	Analysis according to recipient	
National Savings	Purchases or deposits and repayments and withdrawals of each type of savings	Department for National Savings
	Wages and Prices	
Index of Weekly Wage Rates	All industries and services, and for manufacturing industries, with an analysis for men, women and juveniles	Department of Employment
Index of Retail Prices	For all items and for certain groups	Department of Employment
Index numbers of wholesale prices	For certain industrial groups, for certain materials	Department of the Environment Department of Industry

Subject-matter	*Main analyses*	*Source*
Index numbers of agricultural prices	Index numbers for certain livestock and livestock products and cereals and farm crops	Ministry of Agriculture, Fisheries and Food
	Retailing and Catering	
Values of sales by retailers	Index numbers of sales of different kinds of shops, of groups of shops, and of all businesses	Department of Industry

The supplement of the monthly Digest of Statistics. Revised in January of each year, this gives definitions and explanatory notes about the subject matter of the digest. It also gives an index of sources.

Economic trends. Published monthly by the Stationery Office, this gives for each subject a short table and a chart showing how the figures have moved over the last few years. Only the main statistical series relating to employment, output, consumption, prices, trade and finance are dealt with. Cross-references to the *Monthly Digest* are given.

Financial Statistics. This is also published monthly by the Stationery Office. It is prepared by the Central Statistical Office, in collaboration with the Statistics Divisions of Government Departments and the Bank of England. It brings together the key monetary statistics of the United Kingdom. The table below indicates some of its contents.

Subject-matter	*Main analyses*	*Source*
	Central Government	
Net receipts and payments by Board of Inland Revenue	Income tax, surtax, capital gains tax, death duties, stamp duties, payments into Consolidated Fund	Board of Inland Revenue
Consolidated Fund	Revenue—Inland Revenue, Customs and Excise, motor vehicle duties; Expenditure—supply services, payments to Northern Ireland	Treasury
	Local Authorities	
Current Account	Receipts—trading surplus, rent, dividends and interest, rates, grants from central Government; Expenditure—education, housing subsidies, debt interest	C.S.O.
	Banking	
London Clearing Banks	Liabilities—capital, reserve funds, deposits; Assets—cash, money at call and short notice, bills discounted, investments, advances to customers	Bankers' Clearing House
	Company Sector	
Appropriation account of companies	Trading profits, rent, income from abroad; dividends on ordinary shares, taxes, undistributed income	C.S.O.

Standard Industrial Classification. This was drawn up to enable the statistics published by the various government departments to be compared. It has been used for the Census of Production, the Census of Distribution, the Census of Population, and is used by the Department of Employment in their analysis figures of employment and unemployment.

The classification is based on industries and not on occupations, and without regard to who owns or operates them; thus, transport services operated under local authorities are included under transport and not local government services. The unit taken is the establishment and is normally the whole of the premises (such as a factory, a farm or a shop) at a particular address, and all activities carried on at that address are included. Where, however, at a single address there are two or more departments engaged in different activities for which separate records are available each department is regarded as a separate establishment and treated accordingly.

The whole field of economic activity is divided up into 27 orders, or major industrial groups. These are distinguished by roman numerals.

The orders are divided into Minimum List headings, denoted by Arabic numerals. There are 181 of these. Gaps have been left in the numbering to allow for any additions that may become necessary at a later date.

For some purposes it may be desirable to obtain statistics for groups of orders. The following schemes are suggested:

Divisions—Eight divisions identified by first digit of Minimum List Heading codes.

Groups—Fourteen groups identified by the first two digits of the Minimum List Heading codes.

Division	*Group*	*Minimum list headings*	*Order*	*Title*
0	00	001–003	I	Agriculture, Forestry, Fishing
1	10	101–109	II	Mining, Quarrying
2–4				Manufacturing
	21–24	211–240	III	Food, Drink and Tobacco
.	.	.	.	.
.	.	.	.	.
.	.	.	.	.
.	.	.	.	.
.	.	.	.	.
.	41–45			Textiles, Leather and Clothing
.	.	411–429	XIII	Textiles
.	.	431–433	XIV	Leather Goods, Fur
.	.	.	.	.
.	.	.	.	.
.	.	.	.	.
.	.	.	.	.
9	90	901–906	XXVII	Public Administration and Defence

Standard regions. Great Britain has been divided into ten regions for purposes of statistical analysis. They coincide with regions used for economic planning. They are as follows:

North, Yorkshire and Humberside, East Midlands, East Anglia, South-East (this includes the Greater London Area), South-West, Wales, West Midlands, North-West, Scotland.

The Index of Retail Prices. This measures the monthly changes in the level of retail prices compared with the level of prices on the 15th January 1974. It is *not* a cost-of-living index, although it is often inaccurately referred to in this way.

The basis for weighting the index is provided by the Family Expenditure Survey. This survey covers all types of private households in Great Britain and Northern Ireland. A sample of about 20,000 households is chosen so that every household in the country has an equal chance of being included in the sample. Each year a new sample is chosen and the households are visited in rotation throughout the year. They are asked to keep detailed records of their expenditure for 14 consecutive days, and to provide interviewers with information covering longer periods for certain payments such as rent, gas, electricity and with information about incomes. About 70 per cent of the households approached co-operate. The use of a three-year average has the advantage of reducing fluctuations from year to year in purchases, particularly of durable goods.

The index covers the whole field of goods and services on which households spend their money except (*a*) income tax; (*b*) National Insurance contributions; (*c*) insurance (other than of buildings); (*d*) subscriptions to societies and funds; (*e*) "pools" and betting; (*f*) fees to doctors, dentists, etc.; (*g*) capital sums for house purchase (including major structural additions). These items are excluded because of the difficulty or impossibility of identifying a unit whose price can be measured.

Prices are collected each month for a large number of items. It is only necessary to collect prices for those items in each sub-group which account for the greater part of the total expenditure of the sub-group and which together reflect price changes in that sub-group as a whole. Thus, one of the sub-groups in the durable household-goods group is chinaware, glassware, etc. The weight for this sub-group (for 1966) is 2 out of a total of 57 for the group. For this sub-group three items have been chosen; cup and saucer or half tea service, a glass tumbler, and a glass ovenware pie dish or pie plate. However, the weight of 2 was obtained, because on the average households spend £2 on chinaware, etc. out of every £57 they spent on durable household goods, and not merely on the three items for which prices were obtained. The change in prices of these three indicators is representative of, and therefore covers, the price changes for all kinds of chinaware and glassware.

The weighing pattern of the Index of Retail Prices is revised annually in January on the basis of information obtained from the Family Expenditure Survey for the three years ending in the previous June.

The weights used for 1975 are as follows:

I	Food	232
II	Alcoholic drink	82
III	Tobacco	46
IV	Housing	108
V	Fuel and light	53
VI	Durable household goods	70
VII	Clothing and footwear	89
VIII	Transport and vehicles	149
IX	Miscellaneous goods	71
X	Services	52
XI	Meals bought and consumed outside the home	48
	Total	1,000

Each of these groups is divided into sub-groups, *e.g.* transport and vehicles is divided into the following: motoring and cycling and fares.

Since the weights are changed annually the chain base method of computation of the index numbers must be used. Weighted arithmetic averages of price relatives are obtained in respect of each group and for all-items and "linked" back to January 1974 = 100.

The index is calculated in respect of prices ruling on the Tuesday

	Weights 1975	*Index of retail prices (15 January 1974 = 100)* 14 January 1975	*11 November 1975*
Food	232	118·3	141·6
Alcoholic drink	82	118·2	144·5
Tobacco	46	124·0	160·7
Housing	108	110·3	133·8
Fuel and light	53	124·9	161·9
Durable household goods	70	118·3	140·2
Clothing and footwear	89	118·6	130·5
Transport and vehicles	149	130·3	153·4
Miscellaneous goods	71	126·2	147·6
Services	52	115·8	151·6
Meals bought and consumed outside the house	48	118·7	142·1
All items	1,000	119·9	?

Calculate the *"All items"* index for 11 November 1975 (15 January 1974 = 100).

nearest to the middle of each month. Indices are given for each group and sub-group and an all-items index is also computed. They are all given rounded to the nearest first place of decimals and published each month in the *Employment Gazette* and the *Monthly Digest of Statistics*.

	Index of retail prices 11th November 1975 (14th January 1975 = 100)	*Weights 1975*	*Weight × Index*
Food	$\frac{141\cdot6}{118\cdot3} \times 100$	232	27,769
Alcoholic drink	$\frac{144\cdot5}{118\cdot2} \times 100$	82	10,025
Tobacco	$\frac{160\cdot7}{124\cdot0} \times 100$	46	5,961
Housing	$\frac{133\cdot8}{110\cdot3} \times 100$	108	13,101
Fuel and light	$\frac{161\cdot9}{124\cdot9} \times 100$	53	6,870
Durable household goods	$\frac{140\cdot2}{118\cdot3} \times 100$	70	8,296
Clothing and footwear	$\frac{130\cdot5}{118\cdot6} \times 100$	89	9,793
Transport and vehicles	$\frac{153\cdot4}{130\cdot3} \times 100$	149	17,542
Miscellaneous goods	$\frac{147\cdot6}{126\cdot2} \times 100$	71	8,304
Services	$\frac{151\cdot6}{115\cdot8} \times 100$	52	6,808
Meals bought and consumed outside the home	$\frac{142\cdot1}{118\cdot7} \times 100$	48	5,746
All items		1,000	120,215

All items index number 11th November 1975 (14th January 1975 = 100) is 120·215.

All items index number 11th November 1975 (15th January 1974 = 100) is the index number 11th November 1975 (14th January 1975 = 100) multiplied by the index number 14th January 1975 (15th January 1974 = 100) =

$$\frac{120\cdot215 \times 119\cdot9}{100} = 144\cdot2$$

Index numbers of wholesale prices. The base date is 1970. The indices are weighted arithmetic averages of price relatives (the average price in a given month divided by the annual average price in 1970).

The weighting pattern used is the Census of Production of 1968 supplemented by short-term statistics of production. Price indices are given for:

(*a*) materials and fuel purchased by broad sectors of industry, *e.g.*, construction materials, textile industries;
(*b*) the output of broad sectors of industry, *e.g.*, chemical and allied industries, clothing and footwear.

All indices available are published in *Trade and Industry*. A small selection only appears in the *Monthly Digest*.

The Index of Industrial Production. The purpose of this index number is to measure changes in the volume of production in the U.K. The index covers mining and quarrying, manufacturing, industrial construction, gas, electricity and water, but excludes agriculture, trade, transport, finance and all public and private services. It includes repairs.

The index is calculated monthly by the Central Statistical Office in collaboration with the Statistics Division of Government departments.

The index is a weighted arithmetic average of quantity relatives, 1970 being the base year (monthly average in 1970 = 100).

The level of production is expressed as a percentage of the average monthly production in 1970. To ensure comparability between different months, adjustments are made to counteract the effect of calendar months not having the same number of days. No adjustment is made for annual or public holidays, so that monthly changes in the index will show the effect of holidays and other causes of seasonal variation.

Each industry has been given a weight proportional to its net output in 1970, derived from the Census of Production of that year. Most of the quantity relatives are calculated from physical quantities delivered or produced. However, it is not always possible to use physical quantities. In the case of engineering and building industries values are used, but these values are deflated by a suitable price index to take into account changes in prices, so that changes in values measure changes in quantities; in the case of repair work, changes in employment figures are used to indicate changes in quantities. Finally, the quantities of major materials used are sometimes employed to indicate changes in output. An example of this is the case of the item, "Rope, twine and net." The home consumption of hard hemp quantities is used, this being the material used to make rope, twine and net.

The weights given to each industry are based on the values of net output in 1970. The net output of an industry is the selling value of services, finished and partly finished goods produced by that industry during the year less any costs incurred in their production payable to other industries. The estimates of net output are based on the Census of Production figures for 1970.

In addition to the general index, separate index numbers are calculated for individual industrial groups and for certain groups of industries.

Additional indices which have been adjusted for holidays and other causes of seasonal variations are also prepared. These series are designed to eliminate normal month-to-month fluctuations and thus to show the trend more clearly. These indices are given for all industries and for total manufacturing industries.

Manpower. The Department of Employment prepares estimates of the working population. Persons temporarily laid off and part-time workers are counted as full units.

The working population is divided into four groups:

1. Employees in employment (excluding private domestic servants),
2. Employers and self-employed persons,
3. H.M. forces,
4. The registered unemployed.

These estimates are prepared quarterly.

The number of employees in employment in the production industries is estimated monthly and analysed into industries according to the standard industrial classification.

The Index of Basic Weekly Wage Rates. The index provides a measure of the average movement from month to month in the level of full-time basic weekly rates of wages as compared with the level at the 31st July 1972. It does not measure changes in actual earnings. The number of industries and services selected for inclusion in the index is 75, including agriculture, mining, manufacturing, transport, national government service, cinema, theatres, etc. The rates of wages are the minimum or standard rates of wage as fixed by voluntary collective agreements between organisations of employers and workpeople, or arbitration awards, or wages Regulation Orders. The rates are those of manual wage-earners, including shop assistants, but excluding clerical, technical and administrative workers. Rates for time-workers, shift-workers and piece-workers have all been included as appropriate, and the various occupations, both skilled and unskilled, and variations in rates by locality have been taken into account.

The index numbers of the separate industries are combined together to produce index numbers for "all industries" and for "manufacturing industries only," by the use of "weights" approximately proportional to the wages bill in each of the total selected industries in 1970. Thus the index at any date represents relative changes in the weekly wages bill had it been affected only by changes in wage rates, and not by any other factors. Separate indices are computed for men, women, juveniles and all workers.

The Index of Normal Weekly Hours. The index of normal weekly hours measures for the same representative industries and services the average movement from month to month at the level of normal weekly hours compared with the level at 31st July 1972. The normal weekly hours used in the calculation of the index are those in respect of which all rates used in the calculation of the index of weekly rates of wages are payable, and hours for the separate industries are then combined in accordance with their relative importance as measured by the numbers employed at the base date.

The Index of Hourly Rates of Wages. This is the index number of weekly rates of wages divided by the index number of normal weekly hours.

Indices of Volume of Exports and Imports. The volume indices show movements in imports and exports after eliminating variations due to price changes. The formula is:

$$100 \times \frac{\text{Value of goods imported (or exported) in current period at the prices of the chosen base year (1970)}}{\text{Value of goods actually imported (or exported) in the base year}}.$$

The calculation is made quarterly and is based on the quantities and values given in the *Overseas Trade Statistics.* The prices used are the average value per unit calculated from the Accounts.

Import and export unit value index numbers. These index numbers measure the monthly movements in import prices and export prices. They are not called price indices, however, because *unit values* are used and not true prices.

Unit values are obtained by dividing the value of trade of each of a selection of commodities recorded in the month by the corresponding quantities. About 1,000 unit values are calculated.

The unit values for each selection of commodities computed from the *Overseas Trade Statistics* for the current month are multiplied by the quantity weights based on the pattern of trade in 1970. The results are aggregated and compared with the corresponding aggregate of the base year, 1970.

The Balance of Payments. This is an account of the transactions between the residents of a country and non-residents. It also includes the transactions of the government with non-residents, including overseas governments. The account will cover a period of time, and the balance of the account will show the amount owing to the residents and government of a country by the non-residents (a surplus) or the amount owed by a country to non-residents (a deficit) in respect of the transactions for this period.

BALANCE OF PAYMENTS 1973

Current account

Amounts receivable	£m	*Amounts payable*	£m
	A. Visible Trade		
Exports	11,435	Imports	13,810
Visible balance	2,375		
	13,810		13,810
	B. Invisibles		
Private services and transfers (net)	860	Government services and transfers (net)	790
Interest, profits and dividends —private sector (net)	1,290	Interest, profits and dividends —public sector (net)	195
		Invisible balance	1,165
	2,150		2,150
	C. Current balance		
B. Invisible balance	1,165	A. Visible balance	2,375
Current balance	1,210		
	2,375		2,375

Currency flow and official financing

Investment and other capital flows (net)	1,012	Current balance	1,210
Balancing item	408	Total currency flow	210
	1,420		1,420
Balance to be financed	210	Addition to official reserves	210

Net figures are the differences between amounts payable to and amounts payable by the U.K.

1. *Current account.* This covers the imports and exports of goods and services, investment income and transfers.

2. *Government services.* These include military, administrative and diplomatic services.

3. *Government transfers.* These include grants, subscriptions and contributions to international organisations.

4. *Private services.* These include shipping, civil aviation, banking, insurance, films and television and royalties.

5. *Private transfers.* These include cash gifts and the value of goods transferred and services rendered between U.K. private residents and overseas residents without a *quid pro quo.*

6. *The private sector.* This includes public corporations.

7. *Interest, profits and dividends.* This item includes interest on loans and dividends on investments paid abroad or received from abroad.

8. *Investment and other capital flows.* This includes inter-governmental loans, investments by United Kingdom residents overseas and overseas investment in the United Kingdom.

9. *Balancing item.* This figure is the amount necessary to balance the accounts; it represents the net total of errors and omissions in the other items.

10. *Total currency flow.* This is the amount owed to or by the residents (including the government) of the United Kingdom. It is the amount to be settled by gold, currencies and sterling deposits, U.K. treasury bills, commercial bills owned by U.K. banks or residents and lodged with banks, etc., *i.e.* by those various kinds of *transactions with overseas monetary authorities* which will settle a debt.

11. *Addition to official reserves.* The official reserve consists of gold, foreign currencies, special drawing rights and changes in the U.K. reserve position in the I.M.F.

National Income and Expenditure. A *Blue Book* giving considerable information and much detailed analysis is published annually every August by H.M.S.O. Quarterly estimates of National Expenditure, Gross Domestic Formation and the Revenue and Expenditure of Central Government are also published in the *Monthly Digest.*

The National Income is the total incomes of the inhabitants of a country. This is equal to the value of the total production of goods

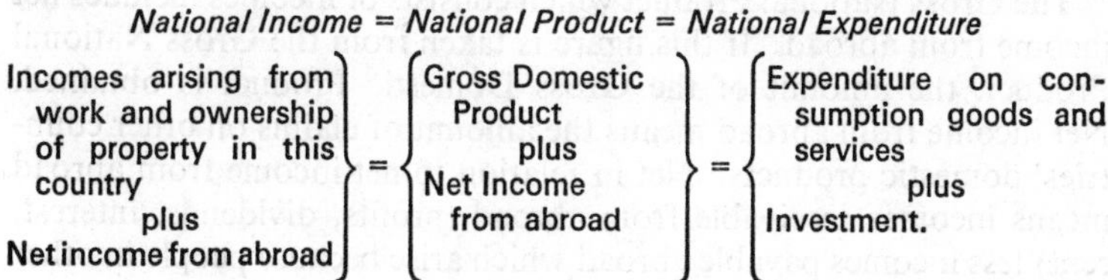

and services whose costs give rise to corresponding incomes. This is equal to the expenditure on consumption goods and services plus capital goods produced (termed investment).

Not all the income of individuals comes into their possession. Some is retained by companies in the form of undistributed profits. Some goes direct to the Government (*e.g.* taxes on profits). These incomes nevertheless form part of the National Income. Some income which individuals receive (*e.g.* National Debt Interest) is in fact income which has been taken in the form of taxes and repaid to other (and sometimes the same) individuals (*e.g.* in the form of interest). This income must not be included twice (*e.g.* once as salary and again as interest). Interest which is not paid out of taxes, but out of the proceeds of the production of goods and services and is part of their cost, is included in the National Income. Incomes paid to the owners of factors of production (labour, capital equipment and land) are costs, and hence total income is equal to the value of total product.

The total payments will give the value of the gross national product at factor cost. It is gross because capital consumption (depreciation) has not been deducted. It is the value of the total goods and services produced. Some of these will be consumed. Other goods will replace existing capital which is being used up or consumed, and some will be an addition to capital (capital formation).

The National Income is Gross National Product less Capital Consumption. Figures for Capital Consumption are given in the *Blue Book*.

The National Product is at factor cost because the incomes to which the payments give rise are to the owners of the factors of production (labour, capital equipment and land) in the form of wages, salaries, rents, interest, dividends, profits, etc. The product, goods and services, will be sold at market prices. The difference consists of expenditure taxes (purchase tax, excise duties, etc.) less subsidies (mainly housing and agricultural) which are received and paid by the Government.

The firms receive payments from consumers for goods and services which they produce. They also receive payments for capital goods. Further, some goods and services are exported. The excess of exports over imports shown is the amount receivable from abroad and is, in effect, part of investment abroad, as is also net income from abroad.

The Gross National Product which consists of incomes includes net income from abroad. If this figure is taken from the Gross National Product, the amount of the Gross Domestic Product is obtained. Net income from abroad means the amount of claims on other countries' domestic products. Net in relation to net income from abroad means incomes receivable from abroad (profits, dividends, interest, rent) less incomes payable abroad which arise because people in other countries own property in this country or have made loans to people in this country.

NATIONAL INCOME AND EXPENDITURE 1973

National Income

Incomes arising from:	£m
Work	49,134
Property (trading profits of companies, trading surpluses of public corporations and trading surpluses of local authorities and central government *and* rent).	13,042
Gross Domestic Product at Factor Cost	62,176
Add Property income from abroad (net)	1,095
Gross National Income at Factor Cost	63,271
Less Capital consumption	7,012
National Income	56,259

National Expenditure

Consumers' expenditure	44,855
Public authorities' current expenditure	13,270
Gross Domestic Capital Formation	14,445
Domestic Expenditure at Market Prices	72,570
Less Disinvestment abroad (see below)	749
Gross National Expenditure at Market Prices	71,821
Less Expenditure taxes (less subsidies)	8,550
Gross National Expenditure at Factor Cost	63,271

National Product

Agriculture and mining	2,744
Manufacturing, construction, gas, electricity and water	25,471
Transport and communication	5,460
Distributive trades	6,122
Insurance, banking, finance, ownership of dwellings	9,442
Public administration, defence, health and education	7,992
Other services	4,945
Gross Domestic Product at Factor Cost	62,176
Add Net property income from abroad	1,095
Gross National Product at Factor Cost	63,271
Imports and income paid abroad	22,291
Less Exports and incomes received from abroad	21,542
	749

Statistical indicators. These are the various economic statistics (price indices, wage indices, index of industrial production, interest rates, balance of payments, etc.) which, when in the form of a time series, indicate the state of the changing economy and give some indication of the direction in which it is going.

Statistics of Trade Act, 1947. This Act provides for the annual holding of a Census of Production and for the carrying out of a Census of Distribution in any year chosen by the Department of Trade and Industry. The Act gives powers to any competent authority to require that any firm shall give information on any matters set out in the schedule to the Act:

> The nature of the undertaking (including its association with other undertakings) and the date of its acquisition; the persons employed or normally employed (including working proprietors), the nature of their employment, their remuneration, and the hours worked; the output, sales, deliveries and services provided; the articles acquired or used, orders, stocks and work-in-progress; the outgoings and costs (including work given out to contractors depreciation, rent, rates and taxes, other than taxes on profits), and capital expenditure; the receipts of and debts owed to the undertaking; the power used or generated; the fixed capital assets, the plant, including the acquisition and disposal of those assets and that plant, and the premises occupied.

Questions

1. Explain how the Index of Weekly Wage Rates is compiled. Explain carefully what it does, and what it does not, measure.

2. Describe the Index of Industrial Production. What are the main differences you will expect to find between movements in this index and variations in the Gross National Product? *Institute of Statisticians.*

3. What information is given in the *Monthly Digest of Statistics* concerning:
(*a*) Wages (*b*) Prices?

4. What are the main problems to be solved in the construction of index numbers? Illustrate by reference to any index numbers of wages or production known to you. *Local Government Examinations Board.*

5. What do the Import and Export Unit Value index numbers measure? How does "unit value" differ from a price?

6. Give an outline of the purpose and method of construction of one of the following index numbers:
(*i*) Index of Retail Prices;
(*ii*) Weekly Wage Rates.
Chartered Institute of Transport.

7. Many of the tables in the *Monthly Digest of Statistics* present information in the form of an index. Name *six* instances, giving the correct description of each series, and name the government department responsible in each case.
What are the main advantages of this kind of presentation, and what are the main problems involved? *Institute of Cost and Management Accountants.*

8. (*a*) Name the sources from which official statistics on the following subjects may be obtained in the United Kingdom:

(*i*) national income and expenditure
(*ii*) employment and unemployment
(*iii*) external trade
(*iv*) population and vital statistics.

(*b*) Describe the information you would expect to find in respect of any two of these subjects and give reasons for the official collection and publication of such statistics. *Association of Certified Accountants.*

CHAPTER 29
STATISTICS AND BUSINESS MANAGEMENT

Budgetary control. Budgets are predetermined plans. The estimates of sales will be the basis of the sales budget, estimates of production, the basis of the production budget, estimates of expenses, the basis of the expense budget, and so on. The budgets will be analysed according to departments, products, areas and any other relevant analysis. Periodic reports will be made, and actual results compared with budgeted plans, which will be continuously revised as further information becomes available and conditions change. The budgets are the basis of control of the activities of a firm. Generally the sales budget is the basis of all the other budgets. For this reason the estimate of future sales is most important. One method is the time series analysis dealt with in Chapter 24. The trend is plotted on graph paper; a smooth curve is drawn between the points (this may, of course, be a straight line); this trend is continued until it reaches the date for which the forecast is being made; the extrapolated trend is multiplied by the appropriate seasonal index; this figure will be the required estimate, subject to any modification called for by any changing conditions. Before the sales figure is adopted for the sales budget, it must be considered from a number of points of view. Perhaps additional plant might be necessary. Would it be possible to finance the budget?

Management ratios. The ratios given below are compiled at suitable intervals. If they are charted on graphs, movements in their value can be noted and the appropriate action taken. It might not be out of place to mention that graphs are used to enable action to be taken, or to refrain from taking action, as the case may be. They are not to be used as pretty pictures.

1. *The 100 per cent statement* (*Balance Sheet*). Each asset is shown as a percentage of the total assets; each liability is shown as a percentage of the total liabilities. Balance Sheets of different undertakings or of the same undertaking at different dates can then be compared.

2. *The 100 per cent statement* (*P. and L. a/c*). The costs shown in the Profit and Loss Account are shown as percentages of net sales. The appropriation of net profit to its various uses can also be given on a percentage basis.

3. *Ratio of Current Assets to Current Liabilities.* This is an extremely

important ratio. Its purpose is to enable a firm to know if it is able to meet its current liabilities. The ratio depends on the rate at which stock is converted into cash. The ratio will vary according to the season.

4. *Stock turnover*. This ratio will measure the rate at which stock is converted into cash or debtors. The formula is:

$$\frac{\text{Cost of goods sold in a given period}}{\text{Average stock of goods during period}}.$$

Alternatively, the turnover ratio may be expressed as the number of days required to sell the stock. This is computed by dividing the number of days in the given period by the ratio as given above. The stock turnover should be computed for every type of commodity. There will be seasonal variations. This ratio measures the efficiency of the buying department.

5. *Debtors' turnover*. This ratio will measure the rate at which debtors are turned into cash. The formula is:

$$\frac{\text{Net credit sales in a given period}}{\text{Average value of debtors during period}}.$$

Alternatively, if the number of days in the given period are divided by the above ratio, the turnover is expressed as the number of days required to collect debts.

Comparison of this figure with the conditions of sale enables the efficiency of the collecting department to be judged.

6. *Ratio of net worth to fixed assets*. A fall in this ratio indicates an expansion of fixed assets, probably financed by borrowing.

7. *Ratio of sales to fixed assets*. This ratio will enable a firm to know if an increase in fixed plant is producing correspondly greater sales.

8. *Ratio of net worth to liabilities*. The ratio of net worth, *i.e.* shareholders' capital plus accumulated undistributed profits to liabilities indicates to what extent creditors are financing the business. If the ratio were 4, this would indicate four-fifths of the capital was provided by the shareholders and one-fifth by the creditors.

9. *Ratio of net profit to net worth*. This ratio indicates whether the funds invested in the business are earning a reasonable return.

10. *Labour turnover*. The cost of replacing workers who leave is high. Labour turnover must therefore be kept at a minimum. A high labour turnover may indicate bad management, discontent or inefficient recruitment. The labour turnover ratio must therefore be continually reviewed. The two principal formulae are:

$$\frac{\text{Number of workers replaced during given period}}{\text{Average number of workers during period}}.$$

or

$$\frac{\text{Number of workers who leave in given period}}{\text{Average number of workers during period}}.$$

The second formula enables the ratio to be analysed into:

$$\frac{\text{Resignations}}{\text{Av. no. workers}}+\frac{\text{Discharges}}{\text{Av. no. workers}}+\frac{\text{Laid-off}}{\text{Av. no. workers}}.$$

Labour turnover may be analysed in many ways, *e.g.* between preventable and non-preventable turnover, or between old employees and new employees.

Internal audit. In the same way as outside auditors check the entries in the books of a firm with the corresponding vouchers, so will staff engaged on an internal audit. The great advantage of an internal audit is that it can be continuous and errors can be put right immediately. It is not necessary to check all entries and vouchers, provided the sample which is checked is truly representative of the whole. Unfortunately, the current practice, both in outside audits and internal audits, is to choose haphazard samples selected according to the whim of the individual auditor, thus leading to biased samples. The sample should be chosen according to the rules of "random sampling" as described in Chapter 3 of this book.

Business reports. These may be:

1. *On a specific subject*, *e.g.* the desirability of installing machine accounting, or, perhaps, an inquiry into the number of strikes in the works in the past year. In this case, the report will follow the usual lines—terms of reference, followed by the body of the report, then the conclusions or findings, followed, if required, by recommendations. Statistical matter often forms part of such reports, *e.g.* tables, charts and graphs.

2. *Periodical reports on certain aspects of the business*, *e.g.* sales, finance, production, advertising, etc. A prominent part of such reports will consist of relevant tabulations with appropriate commentary. These reports form the basis for action and policy decisions. Management ratios will often be shown in such reports.

Purchasing department. Graphs showing stocks of each commodity where the stock held is large, or graphs showing stocks of groups of commodities where each individual item in the group is stocked only in small quantities, will indicate movements in stocks held. When separate commodities are charted, the graphs will show quantities; when groups of commodities are charted, value will have to be shown. On the graphs, distinctive lines will show "ideal" quantity, maximum quantity and minimum quantity or re-ordering quantity. These graphs are extremely useful to prevent overbuying.

Stock turnover figures will indicate if the "right" goods are being purchased.

Graphs of prices over a period of years may indicate the most favourable time of year to purchase, but this may conflict with other considerations: it costs money to hold stocks.

Debt-collection management. A chart of the debtors' turnover will indicate if debts are taking longer to collect. This may be due to accepting less credit-worthy customers or changing terms of sale or less efficient debt management, or possibly changing general economic conditions.

Advertising department. Inquiries and orders should be shown on the same semi-logarithmic chart. Differences between rates of change may indicate inefficient (or otherwise) handling of inquiries. Similarly, advertising costs and sales shown on a semi-logarithmic graph will indicate the effectiveness of advertising. In both cases, there will be a time lag to take into consideration.

Sales department. Tables and/or graphs showing the following information have obvious uses:

(*a*) Comparison of actual and estimated sales.
(*b*) Analysis of sales according to method of sale, *e.g.* mail order, salesmen, dealers, etc.
(*c*) Analysis of sales according to area, commodity, credit or cash, etc.
(*d*) Analysis of sales expenses, advertising, salaries, commission, etc.
(*e*) Analysis of the performance of salesmen, showing calls made, orders obtained, ratio of orders to calls and average cost per order in respect of each salesman.

Production department. Graphs showing production and sales will indicate whether to increase or decrease production. Graphs showing output and labour costs indicate the efficiency of labour. These graphs should be drawn on semi-logarithmic paper.

Among the many charts which are useful mention must be made of Gantt progress charts (see Chapter 10) and quality control charts (see Chapter 26).

Personnel department. Among the matters that may be dealt with by means of tables and charts, the following may be cited:

(*a*) Accidents and injuries, analysed according to seriousness, department and cause.
(*b*) Progress of trainees.
(*c*) Absenteeism analysed according to reason and department.
(*d*) Labour turnover.

Questions

1. Of what use are management ratios? Illustrate your answer by means of examples.

2. You are required to prepare the sales report for your firm. Mention some of the statistical material which will form part of such a report.

3. Explain how time series analysis can be used to prepare the sales budget in budgetary control. What precautions are necessary in fixing the sales budget?

4. Give full explanations of any five of the following technical terms and phrases:

(i) Standard error of a proportion
(ii) Confidence limits
(iii) The Quartiles
(iv) Dispersion
(v) Mean deviation
(vi) Relative error
(vii) Cumulative frequency table
(viii) Standardised deviates

I.C.S.A. Part 1, June 1975.

5. Indicate the more important uses of any TWO of the following charts, and give diagrammatic examples to support your answer:

(*a*) Historigram
(*b*) Lorenz chart
(*c*) Gantt chart
(*d*) Pie diagram

I.C.S.A. Part 1, June 1975.

6. What methods are available to indicate the general change in retail price levels in a community? Support your answer by referring to the construction of statistical indicators in the United Kingdom or a country of your choice.

I.C.S.A. Part 1, June 1975.

7. Explain ANY FIVE of the following technical terms and phrases:

(i) Standard error of the mean
(ii) Fluctuations of sampling
(iii) Random numbers
(iv) Quartile deviation
(v) Coefficient of Variation
(vi) The geometric mean
(vii) Stratified sampling
(viii) Charlier's check

I.C.S.A. Part 1, Dec. 1975.

APPENDIX A

Logarithm Tables

	0	1	2	3	4	5	6	7	8	9	1	2	3	4	5	6	7	8	9
10	0000	0043	0086	0128	0170						5	9	13	17	21	26	30	34	38
						0212	0253	0294	0334	0374	4	8	12	16	20	24	28	32	36
11	0414	0453	0492	0531	0569						4	8	12	16	20	23	27	31	35
						0607	0645	0682	0719	0755	4	7	11	15	18	22	26	29	33
12	0792	0828	0864	0899	0934						3	7	11	14	18	21	25	28	32
						0969	1004	1038	1072	1106	3	7	10	14	17	20	24	27	31
13	1139	1173	1206	1239	1271						3	6	10	13	16	19	23	26	29
						1303	1335	1367	1399	1430	3	7	10	13	16	19	22	25	29
14	1461	1492	1523	1553	1584						3	6	9	12	15	19	22	25	28
						1614	1644	1673	1703	1732	3	6	9	12	14	17	20	23	26
15	1761	1790	1818	1847	1875						3	6	9	11	14	17	20	23	26
						1903	1931	1959	1987	2014	3	6	8	11	14	17	19	22	25
16	2041	2068	2095	2122	2148						3	6	8	11	14	16	19	22	24
						2175	2201	2227	2253	2279	3	5	8	10	13	16	18	21	23
17	2304	2330	2355	2380	2405						3	5	8	10	13	15	18	20	23
						2430	2455	2480	2504	2529	3	5	8	10	12	15	17	20	22
18	2553	2577	2601	2625	2648						2	5	7	9	12	14	17	19	21
						2672	2695	2718	2742	2765	2	4	7	9	11	14	16	18	21
19	2788	2810	2833	2856	2878						2	4	7	9	11	13	16	18	20
						2900	2923	2945	2967	2989	2	4	6	8	11	13	15	17	19
20	3010	3032	3054	3075	3096	3118	3139	3160	3181	3201	2	4	6	8	11	13	15	17	19
21	3222	3243	3263	3284	3304	3324	3345	3365	3385	3404	2	4	6	8	10	12	14	16	18
22	3424	3444	3464	3483	3502	3522	3541	3560	3579	3598	2	4	6	8	10	12	14	15	17
23	3617	3636	3655	3674	3692	3711	3729	3747	3766	3784	2	4	6	7	9	11	13	15	17
24	3802	3820	3838	3856	3874	3892	3909	3927	3945	3962	2	4	5	7	9	11	12	14	16
25	3979	3997	4014	4031	4048	4065	4082	4099	4116	4133	2	3	5	7	9	10	12	14	15
26	4150	4166	4183	4200	4216	4232	4249	4265	4281	4298	2	3	5	7	8	10	11	13	15
27	4314	4330	4346	4362	4378	4393	4409	4425	4440	4456	2	3	5	6	8	9	11	13	14
28	4472	4487	4502	4518	4533	4548	4564	4579	4594	4609	2	3	5	6	8	9	11	12	14
29	4624	4639	4654	4669	4683	4698	4713	4728	4742	4757	1	3	4	6	7	9	10	12	13
30	4771	4786	4800	4814	4829	4843	4857	4871	4886	4900	1	3	4	6	7	9	10	11	13
31	4914	4928	4942	4955	4969	4983	4997	5011	5024	5038	1	3	4	6	7	8	10	11	12
32	5051	5065	5079	5092	5105	5119	5132	5145	5159	5172	1	3	4	5	7	8	9	11	12
33	5185	5198	5211	5224	5237	5250	5263	5276	5289	5302	1	3	4	5	6	8	9	10	12
34	5315	5328	5340	5353	5366	5378	5391	5403	5416	5428	1	3	4	5	6	8	9	10	11
35	5441	5453	5465	5478	5490	5502	5514	5527	5539	5551	1	2	4	5	6	7	9	10	11
36	5563	5575	5587	5599	5611	5623	5635	5647	5658	5670	1	2	4	5	6	7	8	10	11
37	5682	5694	5705	5717	5729	5740	5752	5763	5775	5786	1	2	3	5	6	7	8	9	10
38	5798	5809	5821	5832	5843	5855	5866	5877	5888	5899	1	2	3	5	6	7	8	9	10
39	5911	5922	5933	5944	5955	5966	5977	5988	5999	6010	1	2	3	4	5	7	8	9	10
40	6021	6031	6042	6053	6064	6075	6085	6096	6107	6117	1	2	3	4	5	6	8	9	10
41	6128	6138	6149	6160	6170	6180	6191	6201	6212	6222	1	2	3	4	5	6	7	8	9
42	6232	6243	6253	6263	6274	6284	6294	6304	6314	6325	1	2	3	4	5	6	7	8	9
43	6335	6345	6355	6365	6375	6385	6395	6405	6415	6425	1	2	3	4	5	6	7	8	9
44	6435	6444	6454	6464	6474	6484	6493	6503	6513	6522	1	2	3	4	5	6	7	8	9
45	6532	6542	6551	6561	6571	6580	6590	6599	6609	6618	1	2	3	4	5	6	7	8	9
46	6628	6637	6646	6656	6665	6675	6684	6693	6702	6712	1	2	3	4	5	6	7	7	8
47	6721	6730	6739	6749	6758	6767	6776	6785	6794	6803	1	2	3	4	5	5	6	7	8
48	6812	6821	6830	6839	6848	6857	6866	6875	6884	6893	1	2	3	4	4	5	6	7	8
49	6902	6911	6920	6928	6937	6946	6955	6964	6972	6981	1	2	3	4	4	5	6	7	8

	0	1	2	3	4	5	6	7	8	9	1	2	3	4	5	6	7	8	9
50	6990	6998	7007	7016	7024	7033	7042	7050	7059	7067	1	2	3	3	4	5	6	7	8
51	7076	7084	7093	7101	7110	7118	7126	7135	7143	7152	1	2	3	3	4	5	6	7	8
52	7160	7168	7177	7185	7193	7202	7210	7218	7226	7235	1	2	2	3	4	5	6	7	7
53	7243	7251	7259	7267	7275	7284	7292	7300	7308	7316	1	2	2	3	4	5	6	6	7
54	7324	7332	7340	7348	7356	7364	7372	7380	7388	7396	1	2	2	3	4	5	6	6	7
55	7404	7412	7419	7427	7435	7443	7451	7459	7466	7474	1	2	2	3	4	5	5	6	7
56	7482	7490	7497	7505	7513	7520	7528	7536	7543	7551	1	2	2	3	4	5	5	6	7
57	7559	7566	7574	7582	7589	7597	7604	7612	7619	7627	1	2	2	3	4	5	5	6	7
58	7634	7642	7649	7657	7664	7672	7679	7686	7694	7701	1	1	2	3	4	4	5	6	7
59	7709	7716	7723	7731	7738	7745	7752	7760	7767	7774	1	1	2	3	4	4	5	6	7
60	7782	7789	7796	7803	7810	7818	7825	7832	7839	7846	1	1	2	3	4	4	5	6	6
61	7853	7860	7868	7875	7882	7889	7896	7903	7910	7917	1	1	2	3	4	4	5	6	6
62	7924	7931	7938	7945	7952	7959	7966	7973	7980	7987	1	1	2	3	3	4	5	6	6
63	7993	8000	8007	8014	8021	8028	8035	8041	8048	8055	1	1	2	3	3	4	5	5	6
64	8062	8069	8075	8082	8089	8096	8102	8109	8116	8122	1	1	2	3	3	4	5	5	6
65	8129	8136	8142	8149	8156	8162	8169	8176	8182	8189	1	1	2	3	3	4	5	5	6
66	8195	8202	8209	8215	8222	8228	8235	8241	8248	8254	1	1	2	3	3	4	5	5	6
67	8261	8267	8274	8280	8287	8293	8299	8306	8312	8319	1	1	2	3	3	4	5	5	6
68	8325	8331	8338	8344	8351	8357	8363	8370	8376	8382	1	1	2	3	3	4	4	5	6
69	8388	8395	8401	8407	8414	8420	8426	8432	8439	8445	1	1	2	2	3	4	4	5	6
70	8451	8457	8463	8470	8476	8482	8488	8494	8500	8506	1	1	2	2	3	4	4	5	6
71	8513	8519	8525	8531	8537	8543	8549	8555	8561	8567	1	1	2	2	3	4	4	5	5
72	8573	8579	8585	8591	8597	8603	8609	8615	8621	8627	1	1	2	2	3	4	4	5	5
73	8633	8639	8645	8651	8657	8663	8669	8675	8681	8686	1	1	2	2	3	4	4	5	5
74	8692	8698	8704	8710	8716	8722	8727	8733	8739	8745	1	1	2	2	3	4	4	5	5
75	8751	8756	8762	8768	8774	8779	8785	8791	8797	8802	1	1	2	2	3	3	4	5	5
76	8808	8814	8820	8825	8831	8837	8842	8848	8854	8859	1	1	2	2	3	3	4	5	5
77	8865	8871	8876	8882	8887	8893	8899	8904	8910	8915	1	1	2	2	3	3	4	4	5
78	8921	8927	8932	8938	8943	8949	8954	8960	8965	8971	1	1	2	2	3	3	4	4	5
79	8976	8982	8987	8993	8998	9004	9009	9015	9020	9025	1	1	2	2	3	3	4	4	5
80	9031	9036	9042	9047	9053	9058	9063	9069	9074	9079	1	1	2	2	3	3	4	4	5
81	9085	9090	9096	9101	9106	9112	9117	9122	9128	9133	1	1	2	2	3	3	4	4	5
82	9138	9143	9149	9154	9159	9165	9170	9175	9180	9186	1	1	2	2	3	3	4	4	5
83	9191	9196	9201	9206	9212	9217	9222	9227	9232	9238	1	1	2	2	3	3	4	4	5
84	9243	9248	9253	9258	9263	9269	9274	9279	9284	9289	1	1	2	2	3	3	4	4	5
85	9294	9299	9304	9309	9315	9320	9325	9330	9335	9340	1	1	2	2	3	3	4	4	5
86	9345	9350	9355	9360	9365	9370	9375	9380	9385	9390	1	1	2	2	3	3	4	4	5
87	9395	9400	9405	9410	9415	9420	9425	9430	9435	9440	0	1	1	2	2	3	3	4	4
88	9445	9450	9455	9460	9465	9469	9474	9479	9484	9489	0	1	1	2	2	3	3	4	4
89	9494	9499	9504	9509	9513	9518	9523	9528	9533	9538	0	1	1	2	2	3	3	4	4
90	9542	9547	9552	9557	9562	9566	9571	9576	9581	9586	0	1	1	2	2	3	3	4	4
91	9590	9595	9600	9605	9609	9614	9619	9624	9628	9633	0	1	1	2	2	3	3	4	4
92	9638	9643	9647	9652	9657	9661	9666	9671	9675	9680	0	1	1	2	2	3	3	4	4
93	9685	9689	9694	9699	9703	9708	9713	9717	9722	9727	0	1	1	2	2	3	3	4	4
94	9731	9736	9741	9745	9750	9754	9759	9763	9768	9773	0	1	1	2	2	3	3	4	4
95	9777	9782	9786	9791	9795	9800	9805	9809	9814	9818	0	1	1	2	2	3	3	4	4
96	9823	9827	9832	9836	9841	9845	9850	9854	9859	9863	0	1	1	2	2	3	3	4	4
97	9868	9872	9877	9881	9886	9890	9894	9899	9903	9908	0	1	1	2	2	3	3	4	4
98	9912	9917	9921	9926	9930	9934	9939	9943	9948	9952	0	1	1	2	2	3	3	4	4
99	9956	9961	9965	9969	9974	9978	9983	9987	9991	9996	0	1	1	2	2	3	3	3	4

Given the logarithm to find the number: The logarithm must be looked for in the body of the table (the *mantissa* part only).

APPENDIX B

Table of Squares

	0	1	2	3	4	5	6	7	8	9
10	1 00 00	1 02 01	1 04 04	1 06 09	1 08 16	1 10 25	1 12 36	1 14 49	1 16 64	1 18 81
11	1 21 00	1 23 21	1 25 44	1 27 69	1 29 96	1 32 25	1 34 56	1 36 89	1 39 24	1 41 61
12	1 44 00	1 46 41	1 48 84	1 51 29	1 53 76	1 56 25	1 58 76	1 61 29	1 63 84	1 66 41
13	1 69 00	1 71 61	1 74 24	1 76 89	1 79 56	1 82 25	1 84 96	1 87 69	1 90 44	1 93 21
14	1 96 00	1 98 81	2 01 64	2 04 49	2 07 36	2 10 25	2 13 16	2 16 09	2 19 04	2 22 01
15	2 25 00	2 28 01	2 31 04	2 34 09	2 37 16	2 40 25	2 43 36	2 46 49	2 49 64	2 52 81
16	2 56 00	2 59 21	2 62 44	2 65 69	2 68 96	2 72 25	2 75 56	2 78 89	2 82 24	2 85 61
17	2 89 00	2 92 41	2 95 84	2 99 29	3 02 76	3 06 25	3 09 76	3 13 29	3 16 84	3 20 41
18	3 24 00	3 27 61	3 31 24	3 34 89	3 38 56	3 42 25	3 45 96	3 49 69	3 53 44	3 57 21
19	3 61 00	3 64 81	3 68 64	3 72 49	3 76 36	3 80 25	3 84 16	3 88 09	3 92 04	3 96 01
20	4 00 00	4 04 01	4 08 04	4 12 09	4 16 16	4 20 25	4 24 36	4 28 49	4 32 64	4 36 81
21	4 41 00	4 45 21	4 49 44	4 53 69	4 57 96	4 62 25	4 66 56	4 70 89	4 75 24	4 79 61
22	4 84 00	4 88 41	4 92 84	4 97 29	5 01 76	5 06 25	5 10 76	5 15 29	5 19 84	5 24 41
23	5 29 00	5 33 61	5 38 24	5 42 89	5 47 56	5 52 25	5 56 96	5 61 69	5 66 44	5 71 21
24	5 76 00	5 80 81	5 85 64	5 90 49	5 95 36	6 00 25	6 05 16	6 10 09	6 15 04	6 20 01
25	6 25 00	6 30 01	6 35 04	6 40 09	6 45 16	6 50 25	6 55 36	6 60 49	6 65 64	6 70 81
26	6 76 00	6 81 21	6 86 44	6 91 69	6 96 96	7 02 25	7 07 56	7 12 89	7 18 24	7 23 61
27	7 29 00	7 34 41	7 39 84	7 45 29	7 50 76	7 56 25	7 61 76	7 67 29	7 72 84	7 78 41
28	7 84 00	7 89 61	7 95 24	8 00 89	8 06 56	8 12 25	8 17 96	8 23 69	8 29 44	8 35 21
29	8 41 00	8 46 81	8 52 64	8 58 49	8 64 36	8 70 25	8 76 16	8 82 09	8 88 04	8 94 01
30	9 00 00	9 06 01	9 12 04	9 18 09	9 24 16	9 30 25	9 36 36	9 42 49	9 48 64	9 54 81
31	9 61 00	9 67 21	9 73 44	9 79 69	9 85 96	9 92 25	9 98 56	10 04 89	10 11 24	10 17 61
32	10 24 00	10 30 41	10 36 84	10 43 29	10 49 76	10 56 25	10 62 76	10 69 29	10 75 84	10 82 41
33	10 89 00	10 95 61	11 02 24	11 08 89	11 15 56	11 22 25	11 28 96	11 35 69	11 42 44	11 49 21
34	11 56 00	11 62 81	11 69 64	11 76 49	11 83 36	11 90 25	11 97 16	12 04 09	12 11 04	12 18 01
35	12 25 00	12 32 01	12 39 04	12 46 09	12 53 16	12 60 25	12 67 36	12 74 49	12 81 64	12 88 81
36	12 96 00	13 03 21	13 10 44	13 17 69	13 24 96	13 32 25	13 39 56	13 46 89	13 54 24	13 61 61
37	13 69 00	13 76 41	13 83 84	13 91 29	13 98 76	14 06 25	14 13 76	14 21 29	14 28 84	14 36 41
38	14 44 00	14 51 61	14 59 24	14 66 89	14 74 56	14 82 25	14 89 96	14 97 69	15 05 44	15 13 21
39	15 21 00	15 28 81	15 36 64	15 44 49	15 52 36	15 60 25	15 68 16	15 76 09	15 84 04	15 92 01
40	16 00 00	16 08 01	16 16 04	16 24 09	16 32 16	16 40 25	16 48 36	16 56 49	16 64 64	16 72 81
41	16 81 00	16 89 21	16 97 44	17 05 69	17 13 96	17 22 25	17 30 56	17 38 89	17 47 24	17 55 61
42	17 64 00	17 72 41	17 80 84	17 89 29	17 97 76	18 06 25	18 14 76	18 23 29	18 31 84	18 40 41
43	18 49 00	18 57 61	18 66 24	18 74 89	18 83 56	18 92 25	19 00 96	19 09 69	19 18 44	19 27 21
44	19 36 00	19 44 81	19 53 64	19 62 49	19 71 36	19 80 25	19 89 16	19 98 09	20 07 04	20 16 01
45	20 25 00	20 34 01	20 43 04	20 52 09	20 61 16	20 70 25	20 79 36	20 88 49	20 97 64	21 06 81
46	21 16 00	21 25 21	21 34 44	21 43 69	21 52 96	21 62 25	21 71 56	21 80 89	21 90 24	21 99 61
47	22 09 00	22 18 41	22 27 84	22 37 29	22 46 76	22 56 25	22 65 76	22 75 29	22 84 84	22 94 41
48	23 04 00	23 13 61	23 23 24	23 32 89	23 42 56	23 52 25	23 61 96	23 71 69	23 81 44	23 91 21
49	24 01 00	24 10 81	24 20 64	24 30 49	24 40 36	24 50 25	24 60 16	24 70 09	24 80 04	24 90 01
50	25 00 00	25 10 01	25 20 04	25 30 09	25 40 16	25 50 25	25 60 36	25 70 49	25 80 64	25 90 81
51	26 01 00	26 11 21	26 21 44	26 31 69	26 41 96	26 52 25	26 62 56	26 72 89	26 83 24	26 93 61
52	27 04 00	27 14 41	27 24 84	27 35 29	27 45 76	27 56 25	27 66 76	27 77 29	27 87 84	27 98 41
53	28 09 00	28 19 61	28 30 24	28 40 89	28 51 56	28 62 25	28 72 96	28 83 69	28 94 44	29 05 21
54	29 16 00	29 26 81	29 37 64	29 48 49	29 59 36	29 70 25	29 81 16	29 92 09	30 03 04	30 14 01

	0	1	2	3	4	5	6	7	8	9
55	30 25 00	30 36 01	30 47 04	30 58 09	30 69 16	30 80 25	30 91 36	31 02 49	31 13 64	31 24 81
56	31 36 00	31 47 21	31 58 44	31 69 69	31 80 96	31 92 25	32 03 56	32 14 89	32 26 24	32 37 61
57	32 49 00	32 60 41	32 71 84	32 83 29	32 94 76	33 06 25	33 17 76	33 29 29	33 40 84	33 52 41
58	33 64 00	33 75 61	33 87 24	33 98 89	34 10 56	34 22 25	34 33 96	34 45 69	34 57 44	34 69 21
59	34 81 00	34 92 81	35 04 64	35 16 49	35 28 36	35 40 25	35 52 16	35 64 09	35 76 04	35 88 01
60	36 00 00	36 12 01	36 24 04	36 36 09	36 48 16	36 60 25	36 72 36	36 84 49	36 96 64	37 08 81
61	37 21 00	37 33 21	37 45 44	37 57 69	37 69 96	37 82 25	37 94 56	38 06 89	38 19 24	38 31 61
62	38 44 00	38 56 41	38 68 84	38 81 29	38 93 76	39 06 25	39 18 76	39 31 29	39 43 84	39 56 41
63	39 69 00	39 81 61	39 94 24	40 06 89	40 19 56	40 32 25	40 44 96	40 57 69	40 70 44	40 83 21
64	40 96 00	41 08 81	41 21 64	41 34 49	41 47 36	41 60 25	41 73 16	41 86 09	41 99 04	42 12 01
65	42 25 00	42 38 01	42 51 04	42 64 09	42 77 16	42 90 25	43 03 36	43 16 49	43 29 64	43 42 81
66	43 56 00	43 69 21	43 82 44	43 95 69	44 08 96	44 22 25	44 35 56	44 48 89	44 62 24	44 75 61
67	44 89 00	45 02 41	45 15 84	45 29 29	45 42 76	45 56 25	45 69 76	45 83 29	45 96 84	46 10 41
68	46 24 00	46 37 61	46 51 24	46 64 89	46 78 56	46 92 25	47 05 96	47 19 69	47 33 44	47 47 21
69	47 61 00	47 74 81	47 88 64	48 02 49	48 16 36	48 30 25	48 44 16	48 58 09	48 72 04	48 86 01
70	49 00 00	49 14 01	49 28 04	49 42 09	49 56 16	49 70 25	49 84 36	49 98 49	50 12 64	50 26 81
71	50 41 00	50 55 21	50 69 44	50 83 69	50 97 96	51 12 25	51 26 56	51 40 89	51 55 24	51 69 61
72	51 84 00	51 98 41	52 12 84	52 27 29	52 41 76	52 56 25	52 70 76	52 85 29	52 99 84	53 14 41
73	53 29 00	53 43 61	53 58 24	53 72 89	53 87 56	54 02 25	54 16 96	54 31 69	54 46 44	54 61 21
74	54 76 00	54 90 81	55 05 64	55 20 49	55 35 36	55 50 25	55 65 16	55 80 09	55 95 04	56 10 01
75	56 25 00	56 40 01	56 55 04	56 70 09	56 85 16	57 00 25	57 15 36	57 30 49	57 45 64	57 60 81
76	57 76 00	57 91 21	58 06 44	58 21 69	58 36 96	58 52 25	58 67 56	58 82 89	58 98 24	59 13 61
77	59 29 00	59 44 41	59 59 84	59 75 29	59 90 76	60 06 25	60 21 76	60 37 29	60 52 84	60 68 41
78	60 84 00	60 99 61	61 15 24	61 30 89	61 46 56	61 62 25	61 77 96	61 93 69	62 09 44	62 25 21
79	62 41 00	62 56 81	62 72 64	62 88 49	63 04 36	63 20 25	63 36 16	63 52 09	63 68 04	63 84 01
80	64 00 00	64 16 01	64 32 04	64 48 09	64 64 16	64 80 25	64 96 36	65 12 49	65 28 64	65 44 81
81	65 61 00	65 77 21	65 93 44	66 09 69	66 25 96	66 42 25	66 58 56	66 74 89	66 91 24	67 07 61
82	67 24 00	67 40 41	67 56 84	67 73 29	67 89 76	68 06 25	68 22 76	68 39 29	68 55 84	68 72 41
83	68 89 00	69 05 61	69 22 24	69 38 89	69 55 56	69 72 25	69 88 96	70 05 69	70 22 44	70 39 21
84	70 56 00	70 72 81	70 89 64	71 06 49	71 23 36	71 40 25	71 57 16	71 74 09	71 91 04	72 08 01
85	72 25 00	72 42 01	72 59 04	72 76 09	72 93 16	73 10 25	73 27 36	73 44 49	73 61 64	73 78 81
86	73 96 00	74 13 21	74 30 44	74 47 69	74 64 96	74 82 25	74 99 56	75 16 89	75 34 24	75 51 61
87	75 69 00	75 86 41	76 03 84	76 21 29	76 38 76	76 56 25	76 73 76	76 91 29	77 08 84	77 26 41
88	77 44 00	77 61 61	77 79 24	77 96 89	78 14 56	78 32 25	78 49 96	78 67 69	78 85 44	79 03 21
89	79 21 00	79 38 81	79 56 64	79 74 49	79 92 36	80 10 25	80 28 16	80 46 09	80 64 04	80 82 01
90	81 00 00	81 18 01	81 36 04	81 54 09	81 72 16	81 90 25	82 08 36	82 26 49	82 44 64	82 62 81
91	82 81 00	82 99 21	83 17 44	83 35 69	83 53 96	83 72 25	83 90 56	84 08 89	84 27 24	84 45 61
92	84 64 00	84 82 41	85 00 84	85 19 29	85 37 76	85 56 25	85 74 76	85 93 29	86 11 84	86 30 41
93	86 49 00	86 67 01	86 86 24	87 04 89	87 23 56	87 42 25	87 60 96	87 79 69	87 98 44	88 17 21
94	88 36 00	88 54 81	88 73 64	88 92 49	89 11 36	89 30 25	89 49 16	89 68 09	89 87 04	90 06 01
95	90 25 00	90 44 01	90 63 04	90 82 09	91 01 16	91 20 25	91 39 36	91 58 49	91 77 64	91 96 81
96	92 16 00	92 35 21	92 54 44	92 73 69	92 92 96	93 12 25	93 31 56	93 50 89	93 70 24	93 89 61
97	94 09 00	94 28 41	94 47 84	94 67 29	94 86 76	95 06 25	95 25 76	95 45 29	95 64 84	95 84 41
98	96 04 00	96 23 61	96 43 24	96 62 89	96 82 56	97 02 25	97 21 96	97 41 69	97 61 44	97 81 21
99	98 01 00	98 20 81	98 40 64	98 60 49	98 80 36	99 00 25	99 20 16	99 40 09	99 60 04	99 80 01

Method of Using Table. The first two figures of the number to be squared are found in the first column. The third figure of the number to be squared is found along the top row numbered 0 to 9.

Example. Find the square of 30·7.

The figures 30 will be found in the first column. This will give the row. The figure found in *this* row under column 7 is 94249. This is the square of 307; therefore the square of 30·7 is 942·49.

If the square of a two-figured number *only* is required, the result is given in the column headed "0" *but* the last 2 noughts must then be ignored.

APPENDIX C
Area Under Normal Curve

$$z = \frac{X - \bar{X}}{\sigma}$$

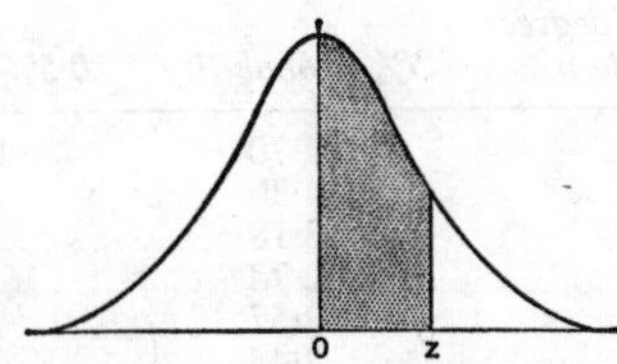

z	0	1	2	3	4	5	6	7	8	9
0·0	0·0000	0·0040	0·0080	0·0120	0·0160	0·0199	0·0239	0·0279	0·0319	0·0359
0·1	0·0398	0·0438	0·0478	0·0517	0·0557	0·0596	0·0636	0·0675	0·0714	0·0754
0·2	0·0793	0·0832	0·0871	0·0910	0·0948	0·0987	0·1026	0·1064	0·1103	0·1141
0·3	0·1179	0·1217	0·1255	0·1293	0·1331	0·1368	0·1406	0·1443	0·1480	0·1517
0·4	0·1554	0·1591	0·1628	0·1664	0·1700	0·1736	0·1772	0·1808	0·1844	0·1879
0·5	0·1915	0·1950	0·1985	0·2019	0·2054	0·2088	0·2123	0·2157	0·2190	0·2224
0·6	0·2258	0·2291	0·2324	0·2357	0·2389	0·2422	0·2454	0·2486	0·2518	0·2549
0·7	0·2580	0·2612	0·2642	0·2673	0·2704	0·2734	0·2764	0·2794	0·2823	0·2852
0·8	0·2881	0·2910	0·2939	0·2967	0·2996	0·3023	0·3051	0·3078	0·3106	0·3133
0·9	0·3159	0·3186	0·3212	0·3238	0·3264	0·3289	0·3315	0·3340	0·3365	0·3389
1·0	0·3413	0·3438	0·3461	0·3485	0·3508	0·3531	0·3554	0·3577	0·3599	0·3621
1·1	0·3643	0·3665	0·3686	0·3708	0·3729	0·3749	0·3770	0·3790	0·3810	0·3830
1·2	0·3849	0·3869	0·3888	0·3907	0·3925	0·3944	0·3962	0·3980	0·3997	0·4015
1·3	0·4032	0·4049	0·4066	0·4082	0·4099	0·4115	0·4131	0·4147	0·4162	0·4177
1·4	0·4192	0·4207	0·4222	0·4236	0·4251	0·4265	0·4279	0·4292	0·4306	0·4319
1·5	0·4332	0·4345	0·4357	0·4370	0·4382	0·4394	0·4406	0·4418	0·4429	0·4441
1·6	0·4452	0·4463	0·4474	0·4484	0·4495	0·4505	0·4515	0·4525	0·4535	0·4545
1·7	0·4554	0·4564	0·4573	0·4582	0·4591	0·4599	0·4608	0·4616	0·4625	0·4633
1·8	0·4641	0·4649	0·4656	0·4664	0·4671	0·4678	0·4686	0·4693	0·4699	0·4706
1·9	0·4713	0·4719	0·4726	0·4732	0·4738	0·4744	0·4750	0·4756	0·4761	0·4767
2·0	0·4772	0·4778	0·4783	0·4788	0·4793	0·4798	0·4803	0·4808	0·4812	0·4817
2·1	0·4821	0·4826	0·4830	0·4834	0·4838	0·4842	0·4846	0·4850	0·4854	0·4857
2·2	0·4861	0·4864	0·4868	0·4871	0·4875	0·4878	0·4881	0·4884	0·4887	0·4890
2·3	0·4893	0·4896	0·4898	0·4901	0·4904	0·4906	0·4909	0·4911	0·4913	0·4916
2·4	0·4918	0·4920	0·4922	0·4925	0·4927	0·4929	0·4931	0·4932	0·4934	0·4936
2·5	0·4938	0·4940	0·4941	0·4943	0·4945	0·4946	0·4948	0·4949	0·4951	0·4952
2·6	0·4953	0·4955	0·4956	0·4957	0·4959	0·4960	0·4961	0·4962	0·4963	0·4964
2·7	0·4965	0·4966	0·4967	0·4968	0·4969	0·4970	0·4971	0·4972	0·4973	0·4974
2·8	0·4974	0·4975	0·4976	0·4977	0·4977	0·4978	0·4979	0·4979	0·4980	0·4981
2·9	0·4981	0·4982	0·4982	0·4983	0·4984	0·4984	0·4985	0·4985	0·4986	0·4986
3·0	0·4987	0·4987	0·4987	0·4988	0·4988	0·4989	0·4989	0·4989	0·4990	0·4990

APPENDIX D

The t Distribution

Number of degrees of freedom	*5% Probability*	*0·5% Probability*
1	12·70	127·00
2	4·30	14·10
3	3·18	7·45
4	2·78	5·60
5	2·57	4·77
6	2·45	4·32
7	2·36	4·03
8	2·31	3·83
9	2·26	3·69

This is, of course, only part of a table. A more complete table would give other probabilities and for other numbers of degrees of freedom.

In dealing with problems of the kind given on page 156 the number of degrees of freedom is one less than the number in the sample. In Example 12 the number in the sample was 5; the number of degrees of freedom was therefore 4. On referring to the above table we see that the probability of getting 5·6 standard errors or more with 4 degrees of freedom is 0·5 per cent.

APPENDIX E

Percentage Points of the χ^2 Distribution

P	*10*	*5*	*2·5*	*1*	*0·5*	*0·1*
d.f. = *1*	2·71	3·84	5·02	6·63	7·88	10·83
2	4·61	5·99	7·38	9·21	10·60	13·81
3	6·25	7·81	9·35	11·34	12·84	16·27
4	7·78	9·49	11·14	13·28	14·86	18·47
5	9·24	11·07	12·83	15·09	16·75	20·52
6	10·64	12·59	14·45	16·81	18·55	22·46
7	12·02	14·07	16·01	18·48	20·28	24·32
8	13·36	15·51	17·53	20·09	21·95	26·12
9	14·68	16·92	19·02	21·67	23·59	27·88
10	15·99	18·31	20·48	23·21	25·19	29·59

This is, of course, only part of a table. A more complete table would give other probabilities and for other numbers of degrees of freedom.

From the table it can be seen that with one degree of freedom the probability of getting χ^2 as great as, or greater than, 3·84 is 5 per cent, a 1 in 20 chance, and this would probably be considered significant.

APPENDIX F

Quality Control Factors

Sample size	*A* Warning	*A* Action	*L*	*M*	*D*
2	1·23	1·94	·18	·80	4·12
3	·67	1·05	·27	·77	2·98
4	·48	·75	·33	·75	2·57
5	·38	·59	·37	·73	2·34
6	·32	·50	·41	·71	2·21

APPENDIX G

Values of e^{-m}

m	1	2	3	4	4·8	5	6
e^{-m}	0·36788	0·13534	0·04979	0·01832	0·00823	0·00674	0·00248

This is only an extract from a much larger table. Values of e^{-m} can easily be calculated by means of logarithms.

Example. Evaluate $e^{-5\cdot5}$.

$e^{-5\cdot5} = \frac{1}{e^{5\cdot5}} = \frac{1}{2\cdot71828^{5\cdot5}}$ ("e" is a constant approximately equal to 2·71828)

No.	*Log*	*Log.Log*
2·71828	0·4343	$\bar{1}$·6378
	5·5	0·7404
$2\cdot71828^{5\cdot5}$	2·389	0·3782
1	0	
0·00409	$\bar{3}$·611	

$e^{-5\cdot5} = 0\cdot00409.$

APPENDIX H

Factorials

x	x!	x	x!	x	x!
1	1	5	120	9	362880
2	2	6	720	10	3628800
3	6	7	5040	11	39916800
4	24	8	40320	12	479001600

$x! = x(x-1)(x-2)(x-3)(x-4)\ldots(3)(2)(1).$
$4! = 4 \times 3 \times 2 \times 1 = 24$

APPENDIX I

Revision Questions

The following questions have been taken from or based on recent examination papers of the following bodies:

The Institute of Chartered Secretaries and Administrators,
The Association of Certified Accountants,
The Institute of Cost and Management Accountants,
The Institute of Company Accountants,
The London Chamber of Commerce and Industry,
The Royal Society of Arts,
The Institute of Municipal Treasurers and Accountants,
The Local Government Examinations Board,
The Institute of Statisticians,
The Institute of Personnel Management,
The Chartered Institute of Transport,
The University of London.

1. Use the method of least squares to forecast the sales of zinc for 1974 from the following time series:

Sales of Zinc by Value (£m)

Year	*Sales of zinc by value*
1969	6·4
1970	6·9
1971	7·3
1972	7·7
1973	8·5

2. (*a*) What is meant by simple random sampling?
(*b*) A random sample of 1,000 manufacturered items is inspected and 250 are found to contain defects. What is the likely range of the proportion defective in the population of items (use 95 per cent confidence limits)?

3. The following table shows the quarterly sales figures in £000s.

Quarterly Sales

Years	*I*	*II*	*III*	*IV*
1970	81	46	42	76
1971	79	40	31	70
1972	64	31	34	66
1973	71	30	32	—

You are required to:

(*a*) use moving averages to calculate the trend;
(*b*) calculate the average seasonal variation;
(*c*) explain how this information can be used in forecasting.

4. What is the importance of the Normal distribution in statistical analysis?

A manufacturer produces tyres which are expected to have a mean life of 24,000 miles (the statistics of tyre life are Normally distributed). If the standard deviation of tyre life is 1,000 miles:

(*a*) What proportion will have worn out after being used for 22,000 miles?

(*b*) What proportion will have been worn out after 25,000 miles of use?

What answers would you have obtained if the mean life had been raised to 24,500 miles?

5. The following table shows the trend of cinema admissions and the growth of television licences in the Sutton Coldfield area during 1950–52:

Year	*Quarter*	*Trend of cinema admissions (thousands)*	*Television licences per 1,000 population*
1950	Third	10,025	24
	Fourth	9,924	37
1951	First	9,814	52
	Second	9,726	64
	Third	9,632	69
	Fourth	9,505	81
1952	First	9,405	98
	Second	9,271	101

Calculate the coefficient of correlation and comment on the result.

6. The following grouped frequency table shows the amounts of mortgages made in a period of six months:

Value of Mortgage Arranged—6 Months Ending 31st Dec. 1973

Value of Mortgage (in £'000s)	*Number of Mortgages arranged*
Under 5	36
5 and up to but not including 10	57
10 " " 15	44
15 " " 20	21
20 " " 25	6

Calculate the mean value of the mortgage and the standard deviation (answers should be given correct to the nearest £1). Use all suitable checks in the working.

7. In a survey of customers' accounts, the following grouped frequency table was obtained dealing with credit made available to clients (in weeks):

Credit Available to Customers

Credit period provided (weeks)	*Number of customers:*
3	46
4	73
5	117
6	35
7	12
8	7

What is the mean credit period provided? In a similar firm it is found that the mean credit period was $5\frac{1}{2}$ weeks (the standard deviation was 1 week). Is there any significant difference between the mean credit periods provided by the two firms?

8. Define the term "error" in statistics.

It is known that the number 45,000 lies between 44,000 and 46,000 and that the number 72,000 lies between 71,500 and 72,500. Find the absolute error when 45,000 is subtracted from 72,000. Calculate this error as a percentage of the difference.

9.

Production of Paper
Weekly Averages (thousand tons)
Newsprint

Quarters	*1*	*2*	*3*	*4*
1969	10·1	9·7	10·0	10·8
1970	11·3	11·2	8·2	10·8
1971	11·9	12·1	10·4	12·1

Calculate the trend of newsprint, together with the seasonal variations.

10. The following data concerning industrial accidents and the absences which result from them involve classification by type of employee:

	Type of employee		
	Men	*Women*	*Juveniles*
Absence following accidents:			
Up to one month	26	16	8
One month or longer	14	9	7

Is there any evidence to suggest that the severity of accident is associated with type of employee?

Table of Chi-Square

No. of degrees of freedom	*5% significance point*	*1% significance point*
2	5·991	9·210
3	7·815	11·345
4	9·488	13·277
5	11·070	15·086
6	12·592	16·812

11. What is the importance of the Normal distribution for the application of quality control methods?

A chocolate manufacturer makes bars of chocolate which have a mean weight of 115 grams and a standard deviation of 12 grams. The manufacturer advertises that the minimum weight is 105 grams. What proportion of bars is likely to be less than this advertised minimum weight?

The manufacturer subsequently decided to advertise that the minimum weight of a bar will be raised to 110 grams. If the standard deviation remains at 12 grams, what mean weight must be obtained to make sure that only 1 per cent of the bars are less than the advertised minimum weight of 110 grams? Assume that the weights of all bars are distributed Normally.

12. Two commodities are of equal importance. Their prices in 1971 and 1972 are given.

	1971	*1972*
Commodity A (£ per ton)	12	15
Commodity B (pence per yd)	5	7

Work out an unweighted price index for 1972 taking 1971 as base. Then work out an unweighted index for 1971 taking 1972 as base. Find the product of the *two* index numbers and criticise your result.

13. Explain what is meant by seasonal variation in a time series. Calculate this measure for the production of cigarette tissue paper from the data given in the table below.

Production of Cigarette Tissue Paper
Weekly averages (tons)

Quarters	*1*	*2*	*3*	*4*
1969	144	142	129	142
1970	138	101	86	102
1971	127	118	108	129

14. "Perfect accuracy is very seldom obtained in statistics." Explain the meaning of this statement. The productivity of 2,675,000 acres is

between 38 and 39 quarters of wheat per acre. A quarter weighs between 470 and 490 lb. Find the yield of wheat in tons.

15. What is meant by the term "the standard error of a proportion observed from a random sample"? In a recent survey, a random sample of adults, including 10,000 persons, provided 62% who thought that cigarette smoking affected health. Calculate the standard error of this proportion, and thus indicate the percentage of adults in the population who took this view, showing a range with 95 per cent confidence limits. What sample size would have been necessary to halve these limits?

16. In order to test the effectiveness of a drying agent in paint, the following experiment was carried out. Each of six samples of material was cut into two halves. One half of each was covered with paint containing the agent and the other half with paint without the agent. Then all twelve halves were left to dry. The time taken to dry was as follows:

Drying time (hours)

	sample number					
	1	2	3	4	5	6
Paint with the agent	3·4	3·8	4·2	4·1	3·5	4·7
Paint without the agent	3·6	3·8	4·3	4·3	3·6	4·7

Required:
Carry out a "t"-test to determine whether the drying agent is effective, giving your reasons for choosing a one-tailed or two-tailed test. Carefully explain your conclusions.

Selected values of $t_{0\cdot05}$

	Degrees of freedom									
	4	5	6	7	8	9	10	11	12	13
One-tailed test	2·13	2·02	1·94	1·90	1·86	1·83	1·81	1·80	1·78	1·77
Two-tailed test	2·78	2·57	2·45	2·36	2·31	2·26	2·23	2·20	2·18	2·16

17. The breaking strength of 80 test pieces of a certain alloy is given in the following table, the unit being given to the nearest thousand pounds per square inch:

Breaking strength	44–46	46–48	48–50	50–52	52–54	Total
No. of pieces	3	24	27	21	5	80

Calculate the average breaking strength of the alloy and the standard deviation. The numbers are distributed uniformly over the interval in which they lie. Using this fact, calculate the percentage of observations lying within the limits of mean $\pm\sigma$, mean $\pm 2\sigma$, mean $\pm 3\sigma$, where σ stands for the standard deviation.

18. The following data show the ranking of 10 bloxite producing factories in descending order of safety consciousness, together with the number of accidents per thousand employees over the last five years.

Factory	*Rank of safety consciousness*	*Number of accidents per 1,000 employees*
A	1	1·2
B	2	5·3
C	3	8·3
D	4	13·1
E	5	18·2
F	6	15·9
G	7	10·4
H	8	3·3
I	9	23·2
J	10	19·7

Required:

(1) By graphing the data, comment on whether accident incidence is associated with safety consciousness.

(2) Calculate Spearman's rank correlation coefficient, R, where:

$$R = 1 - \frac{6\Sigma d^2}{n(n^2 - 1)},$$ and interpret the result.

(3) Now suppose that it is known that factory H is not typical, being much more modern than the others. Without carrying out the calculations, explain what modifications to your approach in (2) above would be required, and in what way the result is likely to be affected.

19. (*a*) A manufacturer estimates that the proportion defective in one of his standard production lines is 5 per cent. He takes a random sample of 100 items and finds that 8 are defective —what probability can be ascribed to the sample? Can we infer that the proportion defective of 5 per cent is justified on the basis of this sample?

(*b*) In City A, a random sample of males reveals that 75 per cent are non-smokers; 1,000 men were questioned to make up the sample. In City B, a similar random sample of 1,000 men gave a proportion of non-smokers as 79 per cent. Can we conclude that there is a significant difference between the two Cities in the proportion of non-smoking men?

20. The table below relates to the weekly pay (before tax and other deductions) of the manual wage-earners on a company's pay-roll:

	April 1970 Numbers	*April 1970* Total pay (£)	*April 1971* Numbers	*April 1971* Total pay (£)
Men, aged 21 and over	350	2,500	300	4,200
Women, aged 18 and over	400	1,600	1,200	8,000
Youths and boys	150	450	100	560
Girls	100	250	400	1,540
	1,000	4,800	2,000	14,300

You are required to construct an index of weekly earnings based on 1970 showing the rise of earnings for all employees as one figure.

21.

1. Required:

(*a*) Define the standard error of a mean of samples of *n* observations and use the concept to explain briefly why one has more confidence in the mean of a large sample than in the mean of a small one.

(*b*) Show why it is unlikely that a sample of 64 observations with a mean of 5·2 was drawn from a population with a mean of 5·5 and a standard deviation of 0·8.

2. An expensive piece of equipment is known to be utilised for about 20 per cent of the time.

In order to obtain a more accurate assessment of its utilisation the management decides to carry out some activity sampling, by observing the equipment at random instants of time and noting the percentage of occasions on which it is working.

Required:

Determine how many observations will be required to be able to state the utilisation of the equipment within ± 5 per cent with 95 per cent certainty.

22. The following data are available concerning the products of a firm taken at random from four factories, and classified according to level of quality:

Output from Four Factories

Levels of Quality	*Factory* A	*Factory* B	*Factory* C	*Factory* D
	(Number of items)			
Low	9	12	6	3
Average	22	52	30	16
Good	69	136	64	81

Is there any evidence to suggest that the level of quality is significantly associated with particular factories?

Table of Chi-square

Number of degrees of freedom	*5% significance point*	*1% signficance point*
3	7·815	11·345
4	9·488	13·277
5	11·070	15·086
6	12·592	16·812
7	14·067	18·475
8	15·507	20·090

23. The following figures, relating to a homogeneous product, were compiled from a company's records:

Number of days taken to complete order	*Number of orders*
0–4	10
5–9	48
10–14	61
15–19	73
20–24	65
25–29	49
30–34	37
35–39	21
40–44	12
45–49	8
50–54	7
55–59	6
60–64	5
65 and over	10

Calculate: (*a*) the median; (*b*) the two quartiles.

How will the results of your calculations help in determining the delivery dates to be quoted on future orders?

24. Briefly explain how the χ^2 (chi-squared) test may be used to decide whether the two characteristics of classification in a contingency table are associated.

Do the following data give any indication that attending a co-educational school is related to degree examination results?

Type of school	*Degree results*		
	Successful	*Failed*	*Total*
Single sex	863	55	918
Co-educational	360	25	385
Total	1,223	80	1,303

Source: Barnett and Lewis, *Journal of the Royal Statistical Society*, 1963.

25. State briefly what you understand by "skewness." For the distribution below calculate the mean, the standard deviation and the median:

Steel rings sold

Diameter (cm)	*Number (gross)*
14·00–14·49	90
14·50–14·99	180
15·00–15·49	200
15·50–15·99	250
16·00–16·49	120
16·50–16·99	80
17·00–17·49	60
17·50–17·99	20

26. In each of the following the degree of approximation in the data is indicated. Obtain in each case the answer required giving the limits of error in the result:

(*a*) Output: 257 million tons; number employed (000) 1,032. Find the output per person.

(*b*) Acreage (000) 1,766; yield per acre 19 cwt. Calculate total output.

27. The average number of employees in a certain firm (excluding staff) was 6,357 in 1953, of whom 43 per cent were absent from their employment for some period during the year. If absences for other than medical reasons are excluded the percentage would be 27. Absence for medical reasons included 187 persons who were absent for periods exceeding two weeks, of whom 15 per cent were absent for more than one month. The number of administrative staff was 628. The exact figure of absentees is not available but is believed to be about 240. There were 176 absentees for duly certified medical reasons, of whom 11 per cent were absent for more than two weeks but less than one month, and 5 persons who were away for periods exceeding one month.

Tabulate these data in a form suitable for presentation in a report to the Board of Directors.

28. The total amount paid in wages by a certain company during 1955 was £W. Let p_i be the per cent of W received by n_i employees who earned £x_i each during the year, so that $p_i = 100n_ix_i/W$. With this notation the distribution of earnings was as follows:

x (£)	250–	350–	450–	550–	650–	750–	850–1,000
p (%)	14	22	29	16	9	6	4

Find the arithmetic and geometric mean earnings of all employees, and estimate what proportion of employees earned more than £10 a week.

29. The table below shows the number of crimes committed in a sample of local authority areas during a certain period.

Population of local authority (10,000s)	*Number of crimes committed*
5	12
7	9
8	14
11	20
13	13
15	14
16	22
19	21
22	29
23	28

For the above data, calculate: (*a*) the product moment correlation coefficient, and (*b*) a rank correlation coefficient. Indicate how you would interpret these coefficients and which you would regard as the more reliable statistic for purposes of interpreting these data.

INDEX